Philip Lutley Sclater

List of the vertebrated animals living in the gardens of the Zoological Society of London

Philip Lutley Sclater

List of the vertebrated animals living in the gardens of the Zoological Society of London

ISBN/EAN: 9783337228668

Printed in Europe, USA, Canada, Australia, Japan

Cover: Foto ©Andreas Hilbeck / pixelio.de

More available books at **www.hansebooks.com**

LIST

OF

VERTEBRATED ANIMALS

LIVING IN

THE GARDENS

OF THE

ZOOLOGICAL SOCIETY

OF LONDON.

1862.

PRINTED FOR THE SOCIETY,
AND SOLD AT THEIR HOUSE IN HANOVER SQUARE.
LONDON:
MESSRS. LONGMAN, GREEN, LONGMANS, AND ROBERTS,
PATERNOSTER ROW.

PARIS:
M. J. ROTHSCHILD,
14, RUE DE BUCY.

LEIPZIG:
M. J. ROTHSCHILD,
2, QUERSTRASSE.

PREFACE.

This List, which has been drawn up, under my superintendence, by Mr. Louis Fraser, contained, as originally prepared, only the Vertebrated animals living in the Gardens on the 31st of December last, on which day it has been the practice for some years to take an accurate census of the Society's stock. The Council having determined to print it, for the use of the Fellows and other persons who take an interest in the Collection, I have carefully revised the whole, and endeavoured to make it as complete as possible by adding the species received since the beginning of the year. A living collection being liable to perpetual change, it cannot, of course, be expected that a list of this sort can be absolutely correct at any given moment; but I believe that the errors and omissions, as it at present stands, are not very numerous.

It will be observed that the Society's Collection, which is supposed to contain the most extensive series of living animals in existence, embraces about 1450 specimens, illustrating 188 species of Mammals, 409 of Birds, 62 of Reptiles, and 23 of Fishes—altogether 682 species of Vertebrates. There is, besides these, a large series of Invertebrated animals of different classes, kept in the House devoted to Aquaria, which varies much in composition and amount. In the " Scientific Index" to the List of Animals in the Society's Gardens, published in 1844 (which is, I believe, the last classified list issued), the number of species exhibited is given as 335—namely 134 Mammals, 197 Birds, 3 Reptiles, and 1 Fish. This shows that we have made considerable progress since that period.

I may add, that I am solely responsible for the arrangement and scientific nomenclature of this list and for the correct determination of the species. The difficulty, I may almost say the impossibility, in many cases, of ascertaining the specific names of living animals with perfect accuracy, is too well known to render apologies necessary in several cases where I may have failed in effecting this object.

PHILIP LUTLEY SCLATER,
Secretary to the Zoological Society of London.

11 Hanover Square,
June 2nd, 1862.

CONTENTS.

MAMMALIA .... *page* 1
    Quadrumana .. *page* 1   Edentata 22
    Carnivora 4   Rodentia 22
    Artiodactyla 11   Chiroptera 27
    Perissodactyla 20   Marsupialia 27
    Proboscidea 22

AVES 30
    Accipitres 30   Gallinæ 64
    Scansores 37   Struthiones 71
    Fissirostres 48   Grallæ 72
    Passeres 49   Anseres 79
    Columbæ 60

REPTILIA 88
    Testudinata 88   Ophidia 91
    Crocodilia 89   Batrachia 94
    Sauria 90

PISCES 97

LIST OF ANIMALS.

Class MAMMALIA.

Order QUADRUMANA.

Family SIMIIDÆ.

Genus SEMNOPITHECUS.

1. *Semnopithecus entellus* (Linn.). Entellus Monkey.
   Hab. Continental India.
   a. Presented by Charles Ashby, Esq., Oct. 10, 1861.

Genus CERCOPITHECUS.

2. *Cercopithecus lalandii*, Is.-Geoff. Vervet Monkey.
   Hab. Southern Africa.
   a. Presented by — Douglas, Esq., Dec. 28, 1861.

3. *Cercopithecus petaurista* (Schreb.). Lesser White-nosed Monkey.
   Hab. Western Africa.
   a. Presented by Bosville Durant, Esq., Sept. 28, 1860.

4. *Cercopithecus diana* (Linn.). Diana Monkey.
   Hab. Western Africa.
   a. Purchased, May 8, 1862.

Genus MACACUS.

5. *Macacus radiatus* (Shaw). Bonnet Monkey.
   Hab. Continental India.
   a. Purchased, 1853.

SIMIIDÆ.

 b. Presented by Miss Porter, March 22, 1860.
 c. Presented by — Richmond, Esq., Dec. 26, 1860.
 d. Presented by Mrs. Blessly, March 14, 1861.

6. *Macacus cynomolgus* (Linn.). Macaque Monkey.
 Hab. Sumatra.
 a. Presented by Mrs. S. C. Hall, Dec. 2, 1861.
 b. Presented by Capt. Barwood, Aug. 12, 1861.
 c. Presented by H. N. Dupree, Esq., Sept. 16, 1861.
 d. Purchased, Oct. 17, 1861.
 e. Presented by — Marsh, Esq., Nov. 25, 1861.

7. *Macacus erythræus* (Schreb.). Rhesus Monkey.
 Hab. Continental India.
 a. Presented by John Bardsley, Esq., Aug. 22, 1861.
 b. Presented by Lady Sartorius, Nov. 19, 1861.
 c. Presented by — Dyer, Esq., June 3, 1861.
 d. Presented by C. A. Long, Esq., Aug. 10, 1861.
 e. Presented by C. Forster, Esq., Oct. 5, 1861.
 f. Presented by — Brooke, Esq., Oct. 25, 1861.
 g. Presented by Colonel Raines, Oct. 31, 1861.
 h. Born in the Menagerie, Sept. 29, 1860.

8. *Macacus nemestrinus* (Linn.). Pig-tailed Monkey.
 Hab. Java.
 a. Presented by W. Brook, Esq., Oct. 4, 1861.

Genus INUUS.

9. *Inuus sylvanus* (Linn.). Barbary Ape.
 Hab. Northern Africa.
 a. Male. Presented by G. B. Crewe, Esq., Nov. 30, 1861.
 b. Male. Purchased, Dec. 15, 1859.

Genus CYNOCEPHALUS.

10. *Cynocephalus porcarius* (Bodd.). Chacma Baboon.
 Hab. Southern Africa.
 a. Male. Purchased, April 22, 1861.

11. *Cynocephalus hamadryas* (Linn.). Arabian Baboon.
 Hab. Aden.
 a. Male. *b*. Female. Presented by Gordon Sandiman, Esq., June 30, 1860.

12. *Cynocephalus anubis*, F. Cuv.   Anubis Baboon.
    Hab. Western Africa.
    a. Male. Purchased, Nov. 16, 1860.

Family CEBIDÆ.

Genus CEBUS.

13. *Cebus apella* (Linn.).   Brown Capuchin Monkey.
    Hab. Guiana.
    a. Male. Deposited by Mr. Robinson, 1860.

Genus HAPALE.

14. *Hapale jacchus* (Linn.).   Marmozet Monkey.
    Hab. South America.
    a. Presented by Capt. N. J. Edwards, R.N., Oct. 31, 1861.

Family LEMURIDÆ.

Genus LEMUR.

15. *Lemur albimanus*, Geoff.   White-handed Lemur.
    Hab. Madagascar.
    a. Purchased, May 29, 1857.

16. *Lemur nigrifrons*, Geoff.   Black-fronted Lemur.
    Hab. Madagascar.
    a. Purchased, May 29, 1857.
    b. Hybrid between this species and *L. albimanus*. Born in the Menagerie, April 17, 1857.

17. *Lemur*, sp. nov.?
    Hab. Madagascar.
    a. Purchased of Mr. Newby, Oct. 28, 1861.

Genus STENOPS.

18. *Stenops javanicus* (Geoff.).   Javan Loris.
    Hab. Java.
    a. Purchased, May 3, 1861.

Order CARNIVORA.

Family CANIDÆ.

Genus CANIS.

19. *Canis lupus*, Linn.   Wolf.
    Hab. Europe.
    a. Male. b. Female. Born in the Menagerie, May 19, 1859.

20. *Canis simensis*, Rüpp.   Abyssinian Wolf.
    Hab. Abyssinia.
    a. Male.   Purchased, 1848.

21. *Canis pallipes*, Sykes.   Indian Wolf.
    Hab. India.
    a. Male.   Purchased, March 10, 1848.

22. *Canis mesomelas*, Schreb.   Black-backed Jackal.
    Hab. South Africa.
    a. Male.   Presented by Edward Spencer, Esq., Jan. 3, 1851.
    b. Female.   Presented by Dr. Hay, Jan. 15, 1858.

23. *Canis niloticus*, Geoff.   Egyptian Fox.
    Hab. North Africa.
    a. Male.   Presented by J. A. Laing, Esq., July 6, 1860.

24. *Canis azaræ*, Pr. Max.   Azara's Fox.
    Hab. South America.
    a. Female.   Born in the Menagerie, May 17, 1858.

25. *Canis argentatus*, Desm.   Silver Fox.
    Hab. North America.
    a. Male. b. Female. Presented by Wm. G. Smith, Esq., Secretary to the Hudson's Bay Company, Oct. 27, 1854.

26. *Canis zaarensis*, Skjöld.   Fennec Fox.
    Hab. Egypt.
    a. Male. b. Female.   Purchased, Aug. 21, 1858.

Family HYÆNIDÆ.

Genus HYÆNA.

27. *Hyæna crocuta*, Erxl.   Spotted Hyæna.
    Hab. South Africa.
    a. Female.   Presented by the Queen of Portugal, April 18, 1853.

VIVERRIDÆ. 5

  b. Male. Presented by Edmund Gabriel, Esq., H.B.M.'s Commissioner at Loando, Angola, Aug. 22, 1860.

28. *Hyæna brunnea,* Thunb.   Brown Hyæna.
  Hab. South Africa.
  a. Male. Purchased, 1853.

Family VIVERRIDÆ.

Genus PARADOXURUS.

29. *Paradoxurus typus,* F. Cuv.   Common Paradoxure.
  Hab. India.
  a. Presented by E. Lowry, Esq., Sept. 3, 1860.

30. *Paradoxurus pallasii,* Gray.   Pallas's Paradoxure.
  Hab. India.
  a. Presented by Samuel Rawlins, Esq., Sept. 15, 1860.
  b. Presented by J. Duplex, Esq., Dec. 10, 1859.
  c. Presented by John Dawson, Esq., Dec. 1, 1859.

31. *Paradoxurus aureus,* F. Cuv.   Golden Paradoxure.
  Hab. Ceylon.
  a. Purchased, Aug. 6, 1860.

Genus NANDINIA.

32. *Nandinia binotata* (Reinw.).   Two-spotted Paradoxure.
  Hab. Western Africa.
  a. Purchased, March 8, 1861.

Genus RYZÆNA.

33. *Ryzæna zenik* (Gmel.).   Suricate.
  Hab. South Africa.
  a. Female. Purchased, Oct. 5, 1859.

Genus HERPESTES.

34. *Herpestes griseus* (Geoff.).   Grey Ichneumon.
  Hab. India.
  a. Female. Presented by Percy Dodgson, Esq., C.M.Z.S., Aug. 15, 1860.
  b. Presented by Charles Clifton, Esq., F.Z.S., Oct. 1, 1861.
  c. Purchased, Oct. 3, 1859.

VIVERRIDÆ.—FELIDÆ.

35. *Herpestes fasciatus* (Desm.). Banded Ichneumon.
    Hab. Eastern Africa.

    a. Male. Presented by Thomas Hawes, Esq., F.Z.S., Sept. 13, 1858. From the Mozambique.

36. *Herpestes auropunctatus*, Hodgs. Spotted Ichneumon.
    Hab. Bengal.

    a. Male. b. Female. Purchased, Aug. 7, 1860.

Genus GENETTA.

37. *Genetta vulgaris* (Linn.). Common Genet.
    Hab. Southern Europe.

    a, b. Females. Purchased, 1861.
    c. Hybrid between this species and male *Genetta senegalensis*. Born in the Menagerie.
    d. Male. Hybrid between this species and male *Genetta tigrina*. Born in the Menagerie, Oct. 5, 1859.

Genus VIVERRICULA.

38. *Viverricula indica* (Geoff.). Indian Civet Cat.
    Hab. India.

    a. Purchased, May 22, 1861.

Genus VIVERRA.

39. *Viverra civetta* (Schreb.). African Civet Cat.
    Hab. Africa.

    a. Presented by the late King of Portugal, Nov. 15, 1855.
    b. Presented by Edmund Gabriel, Esq., H.B.M.'s Commissioner at Loando, Angola, Aug. 22, 1860.

Family FELIDÆ.

Genus FELIS.

40. *Felis leo*, Linn. Lion.
    Hab. Africa and India.

    a. Female. Presented by Mr. Alderman Finnis, March 27, 1856. From Babylonia.
    b. Male. c. Female. Purchased, Oct. 28, 1859. From South Africa.
    d, e, f. Males. g, h. Females. Deposited by Mr. Jamrach, Sept. 14, 1861.

FELIDÆ. 7

41. *Felis tigris*, Linn.   Tiger.
  Hab. India.
  a. Male. b. Female. Presented by Major Marston, April 3, 1858.
  c. Female. Presented by Michael Scott, Esq., 1862.

42. *Felis onca*, Linn.   Jaguar.
  Hab. South America.
  a. Male. Presented by H.E. W. D. Christie, F.Z.S., Minister to the Argentine Confederation, March 16, 1858.

43. *Felis hernandesii* (Gray).   Mexican Jaguar.
  Hab. Mazatlan.
  a. Female. Presented by Miss Mary Knight, Nov. 28, 1857.
  b. Male. Hybrid between this species and male *Felis onca*, Linn. Born in the Menagerie, June 5, 1861.

44. *Felis leopardus*, Linn.   Leopard.
  Hab. Asia.
  a. Male. Presented by Her Majesty The Queen, Feb. 16, 1860.
  b. Female. Presented by the late King of Portugal, Sept. 24, 1856.

45. *Felis varius* (Gray).   African Leopard.
  Hab. Africa.
  a. Male. Presented by J. H. Drummond Hay, Esq., C.M.Z.S., May 24, 1858. From Morocco.
  b. Female. Presented by His Royal Highness the Duke of Oporto, May 14, 1860.

46. *Felis jubata*, Schreb.   Cheetah.
  Hab. Africa.
  a. Male. Purchased, Nov. 19, 1858. From South Africa.
  b. Female. Purchased, Oct. 14, 1857. From South Africa.

47. *Felis macrocelis*, Temm.   Clouded Tiger.
  Hab. Assam.
  a. Male. Purchased, May 16, 1854.

48. *Felis concolor*, Linn.   Puma.
  Hab. America.
  a. Male. b. Female. Presented by H.E. W. D. Christie, F.Z.S., Minister to the Argentine Confederation, Jan. 6, 1859.
  c. Female. Presented, Aug. 29, 1861.
  d. Female. Born in the Menagerie, Sept. 2, 1861.

49. *Felis serval,* Schreb.   Serval.
   Hab. South Africa.
   a. Female. Purchased, Sept. 26, 1859.

50. *Felis pardalis,* Linn.   Ocelot.
   Hab. Tropical America.
   a. Presented by W. Duncan Stewart, Esq., May 3, 1861.

Family MUSTELIDÆ.

Genus PUTORIUS.

51. *Putorius fœtidus,* Linn.   Polecat.
   Hab. British Islands.
   a. Female. Presented by the Hon. Charles Ellis, F.Z.S., 1861.
   b. Male. Purchased, Jan. 9, 1862.

Genus MUSTELA.

52. *Mustela canadensis,* Schreb.   Canadian or Fisher Marten.
   Hab. North America.
   a. Male. b. Female. Presented by Capt. Herd, Oct. 6, 1860.

53. *Mustela martes,* Linn.   Pine Marten.
   Hab. British Islands.
   a. Male. Purchased, July 3, 1854.
   b. Female. Presented by Richard Stennett, Esq., March 10, 1856.

Genus GRISONIA.

54. *Grisonia vittata* (Schreb.).   Grison.
   Hab. South America.
   a. Male. Purchased, Dec. 24, 1860.

Genus LUTRA.

55. *Lutra vulgaris* (Linn.).   Common Otter.
   Hab. British Islands.
   a. Male. b. Female. Presented by the Marquis of Bath, F.Z.S., June 15, 1861.

Genus MELES.

56. *Meles taxus* (Schreb.). Common Badger.
    Hab. British Islands.
    a. Female. Presented by Sir Samuel Morton Peto, M.P., Jan. 28, 1860.
    b. Male. Presented by J. H. Thrupp, Esq., May 22, 1860.
    c. Female. d, e. Young. Presented by the Duke of Richmond, April 22, 1862.
    f. Presented by the Duke of Richmond, April 16, 1862.

Genus MELLIVORA.

57. *Mellivora ratelus* (Sparrm.). Ratel or Bharsiah.
    Hab. South Africa, and India.
    a. Male. Presented by Capt. Tower, R.N., July 11, 1857.
    b. Female. Presented by E. Wemyss, Esq., Nov. 30, 1861.
    c, d. Indian variety. Presented by Arthur Grote, Esq., C.M.Z.S., April 26, 1862.

Genus GULO.

58. *Gulo luscus* (Linn.). Wolverine or Glutton.
    Hab. Europe.
    a. Male. Presented by George Gillett, Esq., F.Z.S., Dec. 1, 1858.

Family URSIDÆ.

Genus ARCTICTIS.

59. *Arctictis binturong* (Raffl.). Binturong.
    Hab. Sumatra.
    a. Male. Presented by Mrs. Samuel Rawson, Oct. 24, 1855.

Genus PROCYON.

60. *Procyon lotor* (Linn.). Raccoon.
    Hab. North America.
    a, b. Males. Presented by Andrew Arcedeckne, Esq., F.Z.S., April 5, 1852.

Genus NASUA.

61. *Nasua rufa,* Desm. Red Coati.
    Hab. Tropical America.
    a. Presented by Sir Malcolm McGregor, Oct. 20, 1859.

Genus URSUS.

62. *Ursus arctos,* Linn. Brown Bear.
    Hab. Northern Europe.
    a. Male. Purchased, June 1, 1852.
    b. Female. Presented by Lieut.-Col. Hon. W. W. Coke, M.P., F.Z.S.
    c. Male. d. Female. Born in the Menagerie, Dec. 27, 1861.
    e. Presented by T. Broadwood, Esq., F.Z.S., May 10, 1853.
    f, g. Young. Var. *collaris.* Deposited by N. Wood, Esq., F.Z.S., April 9, 1862.

63. *Ursus syriacus,* Ehrenb. Syrian Bear.
    Hab. Persia.
    a. Female. Purchased, April 4, 1851.

64. *Ursus isabellinus,* Horsf. Isabelline Bear.
    Hab. India.
    a. Male. Received in exchange, Aug. 16, 1861.

65. *Ursus americanus,* Pall. Black Bear.
    Hab. North America.
    a. Male. Presented by J. Wingfield Malcolm, Esq., Oct. 5, 1860.
    b. Female. Presented by the Officers of the 71st Light Infantry, Dec. 10, 1854.
    c. Female. Presented by Capt. Herd, Oct. 11, 1861.
    d. Male. Presented by G. N. Faulkner, Esq., Aug. 5, 1861.

66. *Ursus torquatus,* Wagn. Himalayan Bear.
    Hab. Northern India.
    a. Female. Presented by W. H. Russell, Esq., Oct. 7, 1859.

67. *Ursus malayanus,* Raffl. Malayan Bear.
    Hab. Sumatra.
    a. Received in exchange from the Zoological Gardens, Rotterdam, March 28, 1862.

Genus THALASSARCTOS.

68. *Thalassarctos maritimus* (Linn.). Polar Bear.
    Hab. Greenland.
    a. Male. Purchased, 1852.
    b. Female. Purchased, 1846.

Genus MELURSUS.

69. *Melursus labiatus* (Blainv.). Sloth Bear.
    Hab. India.
    a. Male. Presented by Capt. Stanley, Sept. 14, 1861.

Family PHOCIDÆ.

Genus PHOCA.

70. *Phoca vitulina*, Linn. Common Seal.
    Hab. British Islands.

    a, b. Purchased, Aug. 2, 1861.

Order ARTIODACTYLA.

Family CERVIDÆ.

Genus CERVUS.

71. *Cervus canadensis*, Briss. Wapiti Deer.
    Hab. North America.

    a. Male. Born in the Menagerie, Aug. 15, 1856.
    b. Female. Born in the Menagerie, Sept. 17, 1851.
    c. Female. Born in the Menagerie, July 25, 1853.

72. *Cervus barbarus*, Benn. Barbary Deer.
    Hab. North Africa.

    a. Male. Presented by Viscount Hill, F.Z.S., Feb. 29, 1860.
    b. Female. Received in exchange from Viscount Powerscourt, F.Z.S., March 11, 1862.

73. *Cervus wallichii*, Cuv. Persian Deer.
    Hab. Persia.

    a. Male. b. Female. Presented by the Earl Ducie, F.Z.S., March 12, 1856.
    c. Female. Born in the Menagerie, Aug. 23, 1858.
    d. Female. Born in the Menagerie, July 24, 1860.
    e. Female. Born in the Menagerie, Aug. 19, 1860.

74. *Cervus duvaucellii*, Cuv. Barasingha Deer.
    Hab. Himalaya.

    a. Male. Presented by the Babu Rajendra Mullick, July 14, 1857.
    b. Female. Purchased, Oct. 26, 1851.
    c. Female. Born in the Menagerie, July 17, 1858.
    d. Male. Born in the Menagerie, Aug. 26, 1861.

75. *Cervus taëvanus*, Blyth. Formosan Deer.
    Hab. Island of Formosa.

    a. Male. Presented by Robert Swinhoe, Esq., H.B.M.'s Vice-Consul, Formosa, C.M.Z.S., Dec. 9, 1861.

CERVIDÆ.

76. *Cervus sika*, Temm.   Japanese Deer.
Hab. Kanegawa, Japan.

a. Male. b. Female.   Presented by J. Wilks, Esq., July 21, 1860.
c. Female.   Presented by Edward Blyth, Esq., C.M.Z.S., Sept. 9, 1861.
d. Female.   Purchased, June 5, 1861.
e. Male.   Born in the Menagerie, Aug. 31, 1861.

77. *Cervus aristotelis*, Cuv.   Sambur Deer.
Hab. India.

a. Male.   Ceylon.   (*C. unicolor*, H. Smith.)   Presented by Her Majesty The Queen, Feb. 23, 1852.
b. Female.   Born in the Menagerie, 1852.
c. Female.   Born in the Menagerie, 1855.
d. Male.   Born in the Menagerie, Feb. 19, 1861.
e. Female.   Born in the Menagerie, Aug. 5, 1861.

78. *Cervus rusa*, Müller.   Rusa Deer.
Hab. Java.

a. Male.   Born in the Menagerie, Oct. 10, 1857.
b. Female.   Received in exchange, March 1, 1851.

79. *Cervus swinhoii*, Sclater.   Swinhoe's Deer.
Hab. Island of Formosa.

a. Male.   Presented by R. Swinhoe, Esq., C.M.Z.S., H.B.M.'s Vice-Consul, Formosa, April 28, 1862.

80. *Cervus porcinus*, Zimm.   Hog Deer.
Hab. Continental India.

a. Female.   Presented by H.E. Sir George Grey, K.C.B., F.Z.S., Governor of New Zealand, April 11, 1859.
b. Male.   Presented by Viscount Hill, F.Z.S., 1855.
c. Male.   Born in the Menagerie, March 9, 1861.
d. Male.   Born in the Menagerie, Dec. 10, 1861.

81. *Cervus* —— ? ——.
Hab. East Indies.

a. Male.   Purchased, April 17, 1862.

82. *Cervus axis*, Erxl.   Axis Deer.
Hab. Continental India.

a. Male.   Presented by Richard Ansdell, Esq., May 5, 1859.

b. Female. Received in exchange from the Société d'Acclimatation de Paris, Dec. 11, 1860.
c. Female. Presented by William Maudsley, Esq., Aug. 29, 1859.

83. *Cervus mexicanus*, H. Smith. Mexican Deer.
Hab. Mexico.
a. Male. Presented by A. Newton, Esq., F.Z.S. Brought from St. Croix, West Indies, Sept. 3, 1857.
b. Male. Presented by Lady Graham, July 25, 1854.

84. *Cervus* —— ? ——.
Hab. North America.
a. Male. Purchased, October 1861.

Genus CERVULUS.

85. *Cervulus reevesii* (Ogilby). Chinese Muntjac.
Hab. China.
a. Male. Presented by Mrs. Dinnesen, April 10, 1861.

86. *Cervulus muntjac* (Zimm.). Javan Muntjac.
Hab. Java.
a. Male. Received in exchange from the Zoological Gardens, Rotterdam, March 28, 1862.

Family MOSCHIDÆ.

Genus MOSCHUS.

87. *Moschus stanleyanus*, Gray. Stanley Musk Deer.
Hab. India.
a. Male. b, c. Females. Presented by the Hon. J. C. Ellis, Aug. 19, 1861.

Family CAMELIDÆ.

Genus AUCHENIA.

88. *Auchenia huanaco* (Mol.). Huanaco.
Hab. Bolivia.
a. Male. b. Female. Presented by H.E. W. D. Christie, F.Z.S., Minister to the Argentine Confederation, March 16, 1858.

89. *Auchenia pacos* (Linn.). Alpaca.
   Hab. Peru.
   a. Male. Received in exchange, Sept. 13, 1860.
   b. Female. Received in exchange, May 19, 1849.

90. *Auchenia glama* (Linn.). Llama.
   Hab. Peru.
   a. Female. Purchased, Jan. 21, 1861.
   b. Male. Received in exchange from the Royal Menagerie at Windsor, April 28, 1860.

Genus CAMELUS.

91. *Camelus dromedarius*, Linn. Common Camel.
   Hab. Egypt.
   a. Male. Presented by His late Highness Ibrahim Pasha, June 29, 1849.

92. *Camelus bactrianus*, Linn. Bactrian Camel.
   Born in the Crimea, 1855.
   a. Female. Presented by the Corps of Royal Engineers, Nov. 18, 1856.

Family CAMELOPARDALIDÆ.

Genus CAMELOPARDALIS.

93. *Camelopardalis giraffa*, Gm. Giraffe.
   Hab. Kordofan.
   a. Male. Born in the Menagerie, April 23, 1846.
   b. Female. Born in the Menagerie, April 25, 1853.
   c. Female. Born in the Menagerie, May 7, 1855.
   d. Female. Born in the Menagerie, May 20, 1861.

Family BOVIDÆ.

Genus OVIS.

94. *Ovis cycloceros*, Hutton. Punjab Wild Sheep.
   Hab. North-west India.
   a. Male. Presented by Brigadier-General Hearsey, F.Z.S., Aug. 19, 1854.
   b. Female. Presented by Major Bartlett, July 25, 1857.
   c. Female. Born in the Menagerie, May 27, 1858.
   d, e. Females. Born in the Menagerie, May 25, 1859.
   f, g. Males. Born in the Menagerie, June 6, 1861.

BOVIDÆ. 15

95. *Ovis tragelaphus*, Desm.   Aoudad.
    Hab. North Africa.
    a. Male.   Presented by H.E. Sir J. Gaspard Le Marchant, Governor of Malta, March 2, 1861.
    b. Female.   Received in exchange from the Société d'Acclimatation of Paris, July 10, 1861.
    c. Female.   Presented by Her Majesty The Queen, May 17, 1862.
    d. Young.   Born in the Menagerie, April 10, 1862.

96. *Ovis musimon*, Schreb.   Mouflon.
    Hab. Sardinia.
    a. Male.   Born in the Menagerie, May 1859.
    b. Female.   Presented by the Earl of Derby, Feb. 26, 1852.
    c. Male.   Hybrid between this species and male *Ovis cycloceros*, Hutt.   Born in the Menagerie, July 2, 1861.

97. *Ovis aries*, Linn., var.   Four-horned Sheep.
    Hab. South Africa.
    a. Male.   Presented by H.E. Sir George Grey, K.C.B., F.Z.S., Governor of New Zealand, Nov. 1, 1861.

Genus CAPRA.

98. *Capra hircus*, Linn., var.   Domestic Goat.
    Hab. Europe.
    a. Male.   Hybrid between Common Goat and a male of the Angolan variety.

99. *Capra caucasica*, Güldenst.   Caucasian Goat.
    Hab. Caucasus.
    a. Male.   Received in exchange from the Zoological Gardens at Amsterdam, Oct. 8, 1860.

100. *Capra megaceros*, Hutton.   Markhoor.
    Hab. Punjab.
    a. Male.   Presented by Capt. Samuel Browne, 2nd Punjab Cavalry, July 21, 1856.
    b. Male.   Hybrid between this species and Common Goat. Born on Mr. Thomas's Farm, 1861.

Genus CALOTRAGUS.

101. *Calotragus campestris* (Thunb.).   Steinbok Antelope.
    Hab. South Africa.
    a. Male.   Presented by H.E. Sir George Grey, K.C.B., F.Z.S., Governor of New Zealand, May 26, 1861.

Genus CEPHALOPHUS.

102. *Cephalophus pygmæus* (Linn.). Blaubok Antelope.
Hab. South Africa.

  a. Male. Presented by H.E. Sir George Grey, K.C.B., F.Z.S., Governor of New Zealand, May 26, 1861.

103. *Cephalophus maxwelli* (H. Smith). Philantomba Antelope.
Hab. Sierra Leone.

  a. Female. Purchased, July 23, 1861.
  b. Female. Hybrid between this species and male *Cephalophus pygmæus* (Schreb.). Born in the Menagerie, March 22, 1862.

104. *Cephalophus rufilatus,* Gray. Coquetoon Antelope.
Hab. Western Africa.

  a. Purchased, Aug. 27, 1861.
  b. Female. Purchased, Jan. 16, 1862.
  c. Female. Purchased, Aug. 22, 1861.

105. *Cephalophus dorsalis,* Gray. Bay Antelope.
Hab. West Africa.

  a. Female. Purchased, Aug. 27, 1861.

106. *Cephalophus mergens* (Blainv.). Duyker-bok Antelope.
Hab. South Africa.

  a. Presented by H.E. Sir George Grey, K.C.B., F.Z.S., Governor of New Zealand, Nov. 1, 1861.

Genus DAMALIS.

107. *Damalis albifrons* (Burch.). Blessbok Antelope.
Hab. South Africa.

  a. Female. Presented by H.E. Sir George Grey, K.C.B., F.Z.S., Governor of New Zealand, May 26, 1861.

Genus ADENOTA.

108. *Adenota lechée,* Gray. Lechée Antelope.
Hab. South Africa.

  a. Female. Presented by H.E. Sir George Grey, K.C.B., F.Z.S., Governor of New Zealand, March 1, 1859.

BOVIDÆ.

Genus GAZELLA.

109. *Gazella dorcas* (Linn.). Gazelle.
 Hab. Egypt.
 a. Male. Presented by H.E. Sir J. Gaspard Le Marchant, Governor of Malta, March 2, 1861.

110. *Gazella euchore* (Forst.). Springbok Antelope.
 Hab. South Africa.
 a. Male. Purchased, Jan. 30, 1861.

Genus ÆGOCERUS.

111. *Ægocerus niger*, Harris. Sable Antelope.
 Hab. South Africa.
 a. Male. Purchased, Sept. 17, 1861.

Genus BOSELAPHUS.

112. *Boselaphus caama* (Cuv.). Hartebeest.
 Hab. South Africa.
 a. Male. Presented by H.E. Sir George Grey, K.C.B., F.Z.S., Governor of New Zealand, Nov. 1, 1861.

Genus ADDAX.

113. *Addax naso-maculatus* (Licht.). Addax.
 Hab. North Africa.
 a. Male. Presented by H.E. Sir J. Gaspard Le Marchant, Governor of Malta, March 2, 1861.
 b. Male. Deposited by Mr. Jamrach, Oct. 11, 1861.

Genus ORYX.

114. *Oryx leucoryx* (Pall.). Leucoryx.
 Hab. North Africa.
 a. Male. Born in the Menagerie, 1853.
 b. Female. Born in the Menagerie, 1852.
 c. Female. Born in the Menagerie, May 17, 1860.

Genus OREAS.

115. *Oreas canna* (Pall.). Eland.
 Hab. South Africa.
 a. Male. Bred by, and received in exchange from, Viscount Hill, F.Z.S., Nov. 26, 1859.

b. Female. Imported in 1850, and bequeathed to the Society by the late Earl of Derby, Dec. 6, 1851.
c. Female. Presented by H.E. Sir George Grey, K.C.B., F.Z.S., Governor of New Zealand, April 11, 1859.
d. Female. Born in the Menagerie, Aug. 10, 1858.
e. Female. Born in the Menagerie, Sept. 1, 1860.
f. Female. Born in the Menagerie, Dec. 20, 1861.
g. Male. Born in the Menagerie, March 17, 1862.

Genus PORTAX.

116. *Portax picta* (Pall.). Nylghaie.
Hab. India.
a. Male. *b.* Female. Born in the Menagerie, May 25, 1856.

Genus CATOBLEPAS.

117. *Catoblepas gnu* (Gm.). White-tailed Gnu.
Hab. South Africa.
a. Male. Purchased, April 14, 1860.

118. *Catoblepas gorgon* (H. Smith). Brindled Gnu.
Hab. South Africa.
a. Female. Purchased, 1859.

Genus ANTILOPE.

119. *Antilope cervicapra,* Linn. Indian Antelope.
Hab. India.
a. Male. Deposited by Capt. Lash, Jan. 18, 1862.

Genus Bos.

120. *Bos grunniens,* Linn. Yak.
Hab. Tibet.
a. Male. *b.* Female. Presented by C. M. Robison, Esq., Feb. 21, 1861.
c. Female. Hybrid between this species and male Zebu. Born in the Menagerie, Jan. 11, 1862.

121. *Bos indicus,* Linn. Zebu.
Hab. Hindostan.
a. Male. Born in the Menagerie, May 4, 1861.
b. Female. Purchased, 1860.
c. Male. *d.* Female. Presented by Dudley Coutts Majoribanks, Esq., M.P., F.Z.S., May 8, 1856.
e. Male. *f.* Female. Brahmin Cattle. Presented by Her Majesty The Queen, May 17, 1862.

Family SUIDÆ.

Genus DICOTYLES.

122. *Dicotyles tajaçu* (Linn.). Collared Peccary.
    Hab. South America.
    a. Male. Presented by Andrew Arcedeckne, Esq., F.Z.S., Sept. 28, 1860.
    b. Female. Presented by Chief Justice Temple, April 13, 1860.
    c. Female. Presented by W. D. Stuart, Esq., May 31, 1861.

123. *Dicotyles labiatus*, Cuv. White-lipped Peccary.
    Hab. South America.
    a. Male. Received in exchange from the Zoological Gardens, Amsterdam, March 30, 1860.

Genus BABIRUSSA.

124. *Babirussa alfurus*, Less. Babirussa.
    Hab. Celebes.
    a. Male. Received in exchange from the Zoological Gardens, Rotterdam, Nov. 11, 1860.

Genus POTAMOCHŒRUS.

125. *Potamochœrus africanus* (Schreb.). South-African River-Hog.
    Hab. South Africa.
    a. Purchased, June 13, 1857.

126. *Potamochœrus penicillatus*, Gray. West-African River-Hog.
    Hab. West Africa.
    a. Female. Born in the Menagerie, June 4, 1858.

Genus SUS.

127. *Sus scrofa*, Linn. Wild Boar.
    Hab. Europe and North Africa.
    a. Female. Presented by H.H. the Grand Duke of Darmstadt, Oct. 9, 1851.
    b. Male. (Var. *S. barbarus*.) Presented by Capt. Daubeny, Jan. 5, 1860. From Barbary.

128. *Sus pliciceps*, Gray. Masked Pig.
    Hab. Japan.
    a. Male. Deposited by Mr. Bartlett, Jan. 21, 1862.
    b, c. Hybrids between this species and Domestic Pig. Deposited by Mr. Bartlett, 1861.

Genus PHACOCHŒRUS.

129. *Phacochœrus æthiopicus* (Pall.). Wart Hog.
    Hab. South Africa.
    a. Male. Purchased, April 23, 1852.

Family HIPPOPOTAMIDÆ.

Genus HIPPOPOTAMUS.

130. *Hippopotamus amphibius*, Linn. Hippopotamus.
    Hab. Upper Nile.
    a. Male. Presented by the late Viceroy of Egypt, May 25, 1850.
    b. Female. Presented by the late Viceroy of Egypt, July 22, 1854.

Order PERISSODACTYLA.

Family EQUIDÆ.

Genus ASINUS.

131. *Asinus quagga* (Linn.). Quagga.
    Hab. South Africa.
    a. Female. Purchased, March 15, 1851.
    b. Male. Presented by H.E. Sir George Grey, K.C.B., F.Z.S., Governor of New Zealand, Sept. 4, 1858.

132. *Asinus burchellii*, Gray. Burchell's Zebra.
    Hab. South Africa.
    a. Male. Purchased.
    b. Female. Presented by H.E. Sir George Grey, K.C.B., F.Z.S., Governor of New Zealand, May 26, 1861.
    c. Male. Deposited by Mr. Jamrach, June 14, 1861.

133. *Asinus hemippus*, Geoff.   Syrian Wild Ass.
   Hab. Western Asia.
   a. Female.   Presented by the late W. Burkhardt Barker, Esq., Oct. 7, 1854.   From Syria.
   b. Female.   Presented by the Hon. C. A. Murray, C.B., F.Z.S., H.B.M.'s Minister Plenipotentiary to the Court of Saxony. From Persia, March 11, 1859.

134. *Asinus indicus*, Sclater.   Indian Wild Ass.
   Hab. Western India.
   a. Female.   Presented by Sir Erskine Perry, F.Z.S., May 5, 1849.   From Cutch.

135. *Asinus hemionus* (Pall.).   Tibetan Wild Ass.
   Hab. Tibet.
   a. Female.   Presented by Major W. E. Hay, F.Z.S., Oct. 22, 1859.

Family TAPIRIDÆ.

Genus TAPIRUS.

136. *Tapirus terrestris* (Linn.).   American Tapir.
   Hab. South America.
   a. Male.   Received in exchange from the Société d'Acclimatation de Paris, Dec. 11, 1860.
   b. Female.   Presented by the late King of Portugal, F.Z.S., Feb. 14, 1857.

Family HYRACIDÆ.

Genus HYRAX.

137. *Hyrax capensis*, Schreb.   Hyrax.
   Hab. South Africa.
   a, b, c. Presented by H.E. Sir George Grey, K.C.B., F.Z.S., Governor of New Zealand, Jan. 8, 1861.

Family RHINOCEROTIDÆ.

Genus RHINOCEROS.

138. *Rhinoceros unicornis*, Linn.   Rhinoceros.
   Hab. India.
   a. Female.   Purchased, July 17, 1850.

Order PROBOSCIDEA.

Family ELEPHANTIDÆ.

Genus ELEPHAS.

139. *Elephas indicus,* Linn.   Indian Elephant.
Hab. India.

  a. Female.   Purchased, May 1, 1851.

Order EDENTATA.

Family BRADYPODIDÆ.

Genus BRADYPUS.

140. *Bradypus didactylus,* Linn.   Two-toed Sloth.
Hab. Brazil.

  a. Received in exchange, Feb. 23, 1851.

Family DASYPODIDÆ.

Genus DASYPUS.

141. *Dasypus encoubert,* Desm.   Weasel-headed Armadillo.
Hab. South America.

  a, b. Purchased, June 30, 1860.

142. *Dasypus villosus,* Gray.   Hairy Armadillo.
Hab. South America.

  a. Presented by C. Bordas, Esq., Oct. 21, 1857.

Order RODENTIA.

Family SCIURIDÆ.

Genus SCIURUS.

143. *Sciurus cinereus,* Linn.   Grey Squirrel.
Hab. North America.

  a. Male. *b.* Female.   Presented by His Royal Highness The Prince of Wales, Nov. 6, 1861.
  c. Male. *d.* Female.   Presented by Mrs. F. Wood, May 18, 1858.

SCIURIDÆ. 23

144. *Sciurus bicolor,* Sparrm.   Jelerang Squirrel.
Hab. India.

　a. Female.   Purchased, July 6, 1858.

Genus SCIUROPTERUS.

145. *Sciuropterus volucellus* (Pall.).   American Flying Squirrel.
Hab. North America.

　a. Male.   *b.* Female.   Born in the Menagerie, Aug. 18, 1860.

Genus ARCTOMYS.

146. *Arctomys marmotta* (Linn.).   Alpine Marmot.
Hab. Europe.

　a. Male.   *b.* Female.   Presented by S. Stauffere, Esq., Feb. 10, 1859.

147. *Arctomys ludovicianus,* Ord.   Prairie Marmot.
Hab. North America.

　a. Female.   Presented by the Proprietors of "The Field" Newspaper, Dec. 29, 1859.

Genus SPERMOPHILUS.

148. *Spermophilus guttatus,* Rich.   American Souslik.
Hab. North America.

　a. Purchased, July 23, 1859.

Genus MYOXUS.

149. *Myoxus muscardinus* (Linn.).   Common Dormouse.
Hab. British Islands.

　a, b. Presented by G. R. Lake, Esq., Oct. 14, 1861.

Genus CHINCHILLA.

150. *Chinchilla lanigera,* Benn.   Chinchilla.
Hab. Chili.

　a. Male.   Born in the Gardens, May 10, 1859.
　b. Female.   Born in the Gardens, April 29, 1858.

Family JERBOIDÆ.

Genus DIPUS.

151. *Dipus ægyptius* (Hasselq.). Jerboa.
 Hab. Egypt.
   a. Male. Presented by the Hon. Mrs. Stuart, Sept. 9, 1861.
   b. Male. c. Female. Purchased, Jan. 14, 1862.

Family MURIDÆ.

Genus MUS.

152. *Mus barbarus,* Linn. Barbary Mouse.
 Hab. Barbary.
   a, b, c. Presented by M. Pichot, Aug. 1, 1860.

Genus FIBER.

153. *Fiber zibethicus* (Linn.). Musquash.
 Hab. North America.
   a. Female. Presented by Lieut. Thomas H. Archer, 39th Regt., Sept. 3, 1860.
   b. Male. Presented by Capt. David Herd, Oct. 11, 1861.

Genus CASTOR.

154. *Castor canadensis,* Kuhl. Castor.
 Hab. North America.
   a, b. Deposited by the Hudson's Bay Company, Oct. 26, 1861.

Genus MYOPOTAMUS.

155. *Myopotamus coypus* (Mol.). Coypu.
 Hab. South America.
   a, b. Purchased, June 3, 1858.

Family HYSTRICIDÆ.

Genus ERETHIZON.

156. *Erethizon dorsatum* (Linn.). Canadian Porcupine.
 Hab. North America.
   a. Presented by Major Boyd, Jan. 24, 1861.

HYSTRICIDÆ.

Genus ATHERURA.

157. *Atherura africana*, Gray.   African Brush-tailed Porcupine.
   Hab. West Africa.
   a, b. Purchased, July 23, 1860.

Genus HYSTRIX.

158. *Hystrix cristata*, Linn.   Crested Porcupine.
   Hab. Europe and Africa.
   a, b. Presented by H. R. H. Prince Alfred, Nov. 12, 1860. From South Africa.

159. *Hystrix leucura*, Sykes.   White-tailed Porcupine.
   Hab. India.
   a. Purchased, 1855.

160. *Hystrix javanica*, F. Cuv.   Javan Porcupine.
   Hab. Java.
   a. Received in exchange from the Zoological Gardens, Rotterdam, May 2, 1860.
   b, c. Received in exchange from the Zoological Gardens, Amsterdam, May 30, 1860.

Genus DASYPROCTA.

161. *Dasyprocta leporina* (Linn.).   Acouchy.
   Hab. South America.
   a, b. Presented by Alfred Ikin, Esq., Nov. 2, 1861.

162. *Dasyprocta aguti* (Linn.).   Golden Agouti.
   Hab. Guiana.
   a. Male. b. Female. Presented by J. S. Caldwell, Esq., May 30, 1857.
   c. Born in the Menagerie, May 24, 1860.
   d, e. Deposited by the Hon. J. C. Ellis, Aug. 19, 1861.

163. *Dasyprocta cristata*, Desm.   Crested Agouti.
   Hab. South America.
   a. Presented by Capt. M. D. Stewart, Oct. 12, 1861.

Genus CŒLOGENYS.

164. *Cœlogenys paca* (Linn.).   Spotted Cavy.
   Hab. South America.
   a. Purchased, Jan. 11, 1862.

Genus CAVIA.

165. *Cavia caprera*, Linn., var.   Restless Cavy or Guinea Pig.
  Hab. South America.
  a, b, c, &c. Domestic variety.   Bred in the Gardens.

Genus HYDROCHŒRUS.

166. *Hydrochœrus capybara*, Erxl.   Capybara.
  Hab. South America.
  a. Purchased, June 13, 1859.

Family LEPORIDÆ.

Genus LEPUS.

167. *Lepus timidus*, Linn.   Common Hare.
  Hab. British Islands.
  a. Male.   Presented by T. Tyrrell, Esq., Jan. 1, 1858.
  b. Male.   Purchased, July 19, 1861.
  c. Male.   Presented by William Goodwin, Esq., Dec. 1, 1859.
  d, e, f. Leporines.   Purchased, June 25, 1860.   From France.
    (Supposed hybrids between *Lepus timidus*, Linn., and *Lepus cuniculus*, Linn.)
  g, h, i, &c. Between Leporine and the Rabbit.   Bred in the Menagerie, 1861.

168. *Lepus variabilis*, Pall.   Alpine Hare.
  Hab. Scotland.
  a. Male.   Presented by Robert Drummond, Esq., F.Z.S., July 2, 1861.

169. *Lepus cuniculus*, Linn.   Common Rabbit.
  Hab. British Islands.
  a. Male. *b, c, d.* Females.   Himalayan or Black-nosed variety. Purchased, 1859.
  e, f. Males.   Moscow variety.   Deposited by Charles Darwin, Esq., F.Z.S., Sept. 30, 1860.
  g, h, i. Males. *j, k, l.* Females.   Silver-grey variety.   Deposited by Mr. Bartlett, 1861.
  m, n. Females.   Variety from Porto Santo.   Deposited by Charles Darwin, Esq., F.Z.S., 1861.

Order CHIROPTERA.

Family PTEROPODIDÆ.

Genus PTEROPUS.

170. *Pteropus medius*, Temm.   Frugivorous Bat.
    Hab. India.
    a. Purchased, Feb. 19, 1861.
    b. Purchased, Feb. 20, 1862.

Order MARSUPIALIA.

Family DIDELPHYIDÆ.

Genus DIDELPHYS.

171. *Didelphys virginianus*, Shaw.   Virginian Opossum.
    Hab. North America.
    a. Male. Presented by A. Downs, Esq., of Halifax, Nova Scotia, C.M.Z.S., Feb. 28, 1860.

172. *Didelphys crassicaudatus*, Desm.   Thick-tailed Opossum.
    Hab. Rio de la Plata.
    a. Female. Presented by H.E. W. D. Christie, F.Z.S., Minister to the Argentine Confederation, July 5, 1860.

Family DASYURIDÆ.

Genus DASYURUS.

173. *Dasyurus viverrinus*, Geoff.   Viverrine Dasyure.
    Hab. New South Wales.
    a. Male. Purchased, July 12, 1860.

174. *Dasyurus maugæi*, Geoff.   Mauge's Dasyure.
    Hab. Australia.
    a. Male. Presented by H.E. Sir George Grey, K.C.B., F.Z.S., Governor of New Zealand, May 26, 1861.

175. *Dasyurus ursinus* (Harr.).   Ursine Dasyure.
    Hab. Tasmania.
    a. Male. Presented by L. C. Stevenson, Esq., March 24, 1860.
    b. Female. Presented by J. C. Wildash, Esq., Aug. 9, 1861.

Genus THYLACINUS.

176. *Thylacinus cynocephalus*, Harr.   Tasmanian Wolf.
   Hab. Van Diemen's Land.

   a. Male.   Purchased, April 9, 1856.

Family PHALANGISTIDÆ.

Genus PHALANGISTA.

177. *Phalangista vulpina* (Shaw).   Vulpine Phalanger.
   Hab. Australia.

   a. Female.   Purchased, June 28, 1860.
   b. Male.   Presented by C. Dawson, Esq., Aug. 26, 1861.
   c. Male.   Presented by Jas. Selfe, Esq., Dec. 14, 1861.

178. *Phalangista fuliginosa*, Ogilby.   Sooty Phalanger.
   Hab. Australia.

   a. Female.   Purchased, March 13, 1861.

Genus BELIDEUS.

179. *Belideus sciureus* (Geoff.).   Squirrel-like Phalanger.
   Hab. Australia.

   a. Male.   Presented by Charles Hutton, Esq., June 21, 1859.
   b. Female.   Born in the Menagerie, April 29, 1860.

180. *Belideus breviceps* (Waterh.).   Short-headed Phalanger.
   Hab. Australia.

   a, b. Females.   Purchased, Oct. 15, 1861.

Family PHASCOLOMYIDÆ.

Genus PHASCOLOMYS.

181. *Phascolomys wombat*, Pér. et Les.   Wombat.
   Hab. Tasmania.

   a. Presented by F. Eardley Wilmot, Esq., April 20, 1853.

Family MACROPODIDÆ.

Genus MACROPUS.

182. *Macropus rufus* (Desm.).   Red Kangaroo.
   Hab. Australia.

   a. Male.   Purchased, June 10, 1860.
   b. Female.   Purchased, July 23, 1861.

MACROPODIDÆ. 29

183. *Macropus melanops*, Gould. Black-faced Kangaroo.
 Hab. South Australia.
 a. Male. Presented by T. R. Fletcher, Esq., April 29, 1859.
 b, c. Males. Purchased, 1862.

184. *Macropus giganteus*, Shaw. Great Kangaroo.
 Hab. New South Wales.
 a. Male. Presented by Wm. Thompson, Esq., July 21, 1855.
 b. Female. Purchased, May 20, 1859.

Genus HALMATURUS.

185. *Halmaturus bennettii*, Waterh. Bennett's Wallaby.
 Hab. Tasmania.
 a. Male. Presented by Thomas Walker, Esq., F.Z.S., Oct. 29, 1856.

186. *Halmaturus ruficollis* (Desm.). Rufous-necked Wallaby.
 Hab. New South Wales.
 a, b. Females. Presented by Thomas Maynard, Esq., June 8, 1856.
 c. Female. Presented by Jas. Selfe, Esq., Dec. 14, 1861.
 d. Male. Hybrid between this species and *H. bennettii*, Waterh. Born in the Menagerie, 1861.

187. *Halmaturus thetidis* (Less.). Pademeleon Wallaby.
 Hab. Australia.
 a. Male. *b.* Female. Purchased, March 10, 1862.

188. *Halmaturus* ——— ? ———. ——— ? ——— Wallaby.
 Hab. Australia.
 a. Purchased, March 10, 1862.

Class AVES.

Order ACCIPITRES.

Family VULTURIDÆ.

Genus CATHARTES.

1. *Cathartes atratus* (Bartr.). Black Vulture.
   Hab. America.

   a, b. Presented by Dr. H. B. Holbeck, of Charleston, U. S. A., July 25, 1860.

Genus SARCORAMPHUS.

2. *Sarcoramphus gryphus* (Linn.). Condor Vulture.
   Hab. South America.

   a. Male. Purchased, 1853.
   b. Female. Purchased, April 29, 1856.

Genus GYPARCHUS.

3. *Gyparchus papa* (Linn.). King Vulture.
   Hab. Tropical America.

   a. Male. Presented by the late King of Portugal, F.Z.S., Nov. 16, 1856.
   b. Female. Presented by H.E. W. D. Christie, Minister to the Argentine Confederation, May 13, 1857.

Genus GYPAËTUS.

4. *Gypaëtus barbatus* (Linn.). Bearded Vulture.
   Hab. Europe.

   a. Received in exchange from the Jardin des Plantes, Paris, Oct. 4, 1851.

Genus OTOGYPS.

5. *Otogyps auricularis* (Daud.). Sociable Vulture.
   Hab. Africa.

   a. Purchased, March 28, 1850.

Genus VULTUR.

6. *Vultur cinereus,* Linn. Cinereous Vulture.
   Hab. Eastern Europe.
   a. Purchased, Nov. 8, 1851.

Genus GYPS.

7. *Gyps fulvus* (Gm.). Griffon Vulture.
   Hab. Europe.
   a. Presented by Cuthbert Wigham, Esq., July 26, 1856.
   b. Presented by Col. Harding, Sept. 12, 1855. From the Crimea.

Genus SERPENTARIUS.

8. *Serpentarius reptilivorus,* Daud. Secretary Vulture.
   Hab. Africa.
   a. South Africa. Presented by E. L. Layard, Esq., F.Z.S., May 3, 1861.
   b. Eastern Africa. Purchased, June 18, 1858.

Family FALCONIDÆ.

Subfamily POLYBORINÆ.

Genus MILVAGO.

9. *Milvago australis* (Gm.). Forster's Milvago.
   Hab. Falkland Islands.
   a. Purchased, Jan. 6, 1859.

10. *Milvago chimango* (Vieill.). Chimango.
    Hab. South America.
    a. Purchased, Nov. 8, 1851.

Genus POLYBORUS.

11. *Polyborus brasiliensis* (Gm.). Brazilian Caracara.
    Hab. South America.
    a. Presented by Hugh Cuming, Esq., C.M.Z.S., Oct. 10, 1831.

FALCONIDÆ.

Subfamily MILVINÆ.

Genus MILVUS.

12. *Milvus regalis,* Briss.   Common Kite.
    Hab. Europe.
    a. Purchased, Jan. 13, 1860.
    b. Presented by the late King of Portugal, F.Z.S., Aug. 3, 1860.

13. *Milvus parasiticus* (Daud.).   Egyptian Kite.
    Hab. West Africa.
    a. Presented by J. H. Gurney, Esq., M.P., F.Z.S., April 13, 1858.

14. *Milvus niger,* Briss.   Black Kite.
    Hab. Europe and Northern Africa.
    a. Purchased, Sept. 7, 1860.
    b. Purchased, July 12, 1861.
    c. Presented by the Hon. Mrs. Stuart, Dec. 16, 1861. From Lake Menzaleh, Egypt.

15. *Milvus govinda,* Sykes.   Indian Kite.
    Hab. Eastern Asia.
    a. Purchased.
    b. Presented by G. B. Bird, Esq., May 26, 1861. From Northern China.

Subfamily AQUILINÆ.

Genus BUTEO.

16. *Buteo vulgaris,* Bechst.   Common Buzzard.
    Hab. England, Europe, North Africa.
    a. Presented by Wm. Broderick, Esq., May 23, 1861.
    b. Presented by J. Gurney Barclay, Esq., Aug. 7, 1860.

17. *Buteo jacal* (Daud.).   Jackal Buzzard.
    Hab. South Africa.
    a. Purchased, 1858.
    b. Presented by H.E. Sir George Grey, K.C.B., F.Z.S., Nov. 1, 1861.

18. *Buteo tachardus* (Daud.).   African Buzzard.
    Hab. South Africa.
    a. Presented by E. L. Layard, Esq., F.Z.S., Oct. 31, 1860.
    b. Purchased, 1862.

FALCONIDÆ.

Genus ARCHIBUTEO.

19. *Archibuteo lagopus* (Gm.).   Rough-legged Buzzard.
    Hab. Europe.
    a, b. Presented by Sir T. Fowell Buxton, Bart., F.Z.S., March 31, 1862.

Genus PERNIS.

20. *Pernis apivorus* (Linn.).   Honey Buzzard.
    Hab. Europe.
    a. Presented by Charles Clifton, Esq., F.Z.S., Sept. 24, 1860.
    b, c. Presented by Percy S. Godman, Esq., C.M.Z.S. From Norway, Aug. 13, 1861.

Genus HALIASTUR.

21. *Haliastur indus* (Bodd.).   Brahminy Kite.
    Hab. Southern Asia.
    a. India. Presented by the Babu Rajendra Mullick, July 14, 1857.
    b. Java. Received in exchange from the Zoological Gardens, Rotterdam, Aug. 7, 1861.

Genus HELOTARSUS.

22. *Helotarsus ecaudatus* (Shaw).   Bateleur Eagle.
    Hab. South Africa.
    a. Presented by the Queen of Portugal, July 14, 1853.

Genus HALIAËTUS.

23. *Haliaëtus albicilla* (Linn.).   Common Sea Eagle.
    Hab. Europe.
    a. Presented by Viscount Powerscourt, F.Z.S., Feb. 13, 1862.

24. *Haliaëtus leucocephalus* (Linn.).   White-headed Sea Eagle.
    Hab. North America.
    a, b. Presented by Earl Fitzwilliam, Feb. 11, 1858.
    c. Presented by J. Wolcot Lambe, Esq., Aug. 31, 1861. From Vancouver's Island.
    d. Presented by Dr. E. J. Longton. Taken in the Atlantic, Feb. 14, 1862.

Genus GERANOAËTUS.

25. *Geranoaëtus aguia* (Temm.). Chilian Sea Eagle.
    Hab. Chili.
    a. Presented by Admiral Seymour, July 12, 1848.

Genus AQUILA.

26. *Aquila chrysaëtos* (Linn.). Golden Eagle.
    Hab. Europe and North America.
    a. Presented by Capt. David Herd, Aug. 13, 1858. From the Hudson's Bay Territory.
    b. Presented by John Henry Gurney, Esq., F.Z.S., 1857.

27. *Aquila heliaca*, Savig. Imperial Eagle.
    Hab. Europe.
    a. Purchased, March 8, 1861.
    b. Presented by the late King of Portugal, F.Z.S., Aug. 30, 1860.

28. *Aquila nævioïdes*, Cuv. Tawny Eagle.
    Hab. North and South Africa.
    a. Presented by E. L. Layard, Esq., F.Z.S., Oct. 31, 1860. From the Cape Colony.
    b. Presented by W. H. Simpson, Esq., 1857. From Algeria.
    c. Presented by Thomas Newell, Esq., Dec. 26, 1861. From Suez.

29. *Aquila audax* (Lath.). Wedge-tailed Eagle.
    Hab. Australia.
    a, b. Presented by Dr. Müller of Melbourne, C.M.Z.S., Jan. 8, 1861.
    c, d. Presented by Dr. Müller of Melbourne, C.M.Z.S., Dec. 11, 1860.
    e, f. Presented by Dr. Müller of Melbourne, C.M.Z.S., Dec. 13, 1859.
    g. Presented by Samuel Magnus, Esq., May 15, 1858.

Genus SPIZAËTUS.

30. *Spizaëtus occipitalis* (Daud.). Black Crested-Eagle.
    Hab. West Africa.
    a. Presented by Edmund Gabriel, Esq., H.B.M.'s Commissioner at Loando, Angola, Sept. 4, 1860.
    b. Purchased, Sept. 10, 1861.

FALCONIDÆ. 35

Genus Thrasaëtus.

31. *Thrasaëtus harpyia* (Linn.).   Harpy Eagle.
    Hab. South America.
    a. Purchased, April 5, 1854.
    b. Presented by the late King of Portugal, F.Z.S., Oct. 15, 1858.

Subfamily Falconinæ.

Genus Falco.

32. *Falco grœnlandicus*, Hancock.   Greenland Falcon.
    Hab. Greenland.
    a. Purchased, 1859.

33. *Falco peregrinus*, Linn.   Peregrine Falcon.
    Hab. Europe.
    a. Presented by J. H. Gurney, Esq., M.P., F.Z.S., April 13, 1858.
    b. Presented by H. Footet, Esq., July 26, 1860.
    c, d. Presented by Major Magenis, Nov. 8, 1860. From the Hebrides.
    e. Presented by Dr. Bree, Aug. 1, 1861.
    f. Deposited by Mr. Edwards, Sept. 21, 1861.

34. *Falco anatum*, Bonap.   Duck Falcon.
    Hab. North America.
    a. Presented by Capt. Spencer, Jan. 26, 1861.

Genus Hypotriorchis.

35. *Hypotriorchis subbuteo* (Linn.).   Hobby.
    Hab. British Islands.
    a. Purchased, Aug. 21, 1861.

36. *Hypotriorchis æsalon* (Gm.).   Merlin.
    Hab. British Islands.
    a. Purchased, Oct. 2, 1861.

Genus Hieracidea.

37. *Hieracidea berigora*, Vig. et Horsf.   Berigora Hawk.
    Hab. Australia.
    a. Purchased, July 18, 1861.

FALCONIDÆ.—STRIGIDÆ.

Genus TINNUNCULUS.

38. *Tinnunculus alaudarius,* Briss.   Common Kestrel.
    Hab. British Islands.
    a. Presented by A. Dodd, Esq., Sept. 27, 1861.

39. *Tinnunculus sparverius* (Linn.).   American Kestrel.
    Hab. N. America.
    a. Purchased of Mr. Newby, Oct. 28, 1861.

Genus ASTUR.

40. *Astur palumbarius* (Linn.).   Goshawk.
    Hab. Europe.
    a. Deposited by J. Wolf, Esq., F.Z.S., Oct. 11, 1861.

41. *Astur novæ-hollandiæ* (Gm.).   White Goshawk.
    Hab. Australia.
    a. Purchased, July 12, 1859.

42. *Astur monogrammicus,* Temm.   One-streaked Falcon.
    Hab. West Africa.
    a. Purchased, June 7, 1861.

Family STRIGIDÆ.

Genus BUBO.

43. *Bubo maximus* (Aldrov.).   Great Eagle Owl.
    Hab. Europe.
    a. Presented by Sir John Cathcart, Bart., April 28, 1859.
    b. Presented by Lieut.-Gen. C. R. Fox, June 22, 1858.
    c. Presented by Edward Fontaine, Esq., April 13, 1858.

44. *Bubo virginianus* (Gm.).   Virginian Eagle Owl.
    Hab. North America.
    a. Presented by Jos. Radford, Esq., Aug. 17, 1861.
    b. Presented by Capt. Herd, Oct. 13, 1858.
    c. Presented by W. G. Smith, Esq., Secretary to the Hudson's Bay Company, Oct. 26, 1854.
    d, e. Presented by Capt. Wishart, Oct. 12, 1858.
    f. Presented by W. G. Smith, Esq., Secretary to the Hudson's Bay Company, Oct. 26, 1858.

45. *Bubo lacteus* (Temm.).   Milky Owl.
    Hab. West Africa.
    a. Received in exchange from the Zoological Gardens, Antwerp, Dec. 13, 1861.

Genus BRACHYOTUS.

46. *Brachyotus palustris* (Bonap.).   Short-eared Owl.
    Hab. Europe.
    a. Purchased, Nov. 22, 1860.

Genus NYCTEA.

47. *Nyctea nivea* (Daud.).   Snowy Owl.
    Hab. Northern Europe.
    a. Purchased, 1858. From Shetland.
    b. Presented by George Clive, Esq., Oct. 11, 1860. From Ireland.

Genus SYRNIUM.

48. *Syrnium aluco* (Linn.).   Wood Owl.
    Hab. Europe.
    a, b. Presented by Edward Newton, Esq., F.Z.S., May 28, 1851. From Norway.
    c, d. Presented by Percy S. Godman, Esq., C.M.Z.S., Aug. 13, 1861. From Norway.

Genus STRIX.

49. *Strix flammea*, Linn.   Common Barn Owl.
    Hab. British Islands.
    a. Presented by Mrs. Smale, Sept. 19, 1856.

50. *Strix personata*, Vig.   Australian Barn Owl.
    Hab. Australia.
    a. Presented by Thomas Woolley, Esq., F.Z.S., June 11, 1857.

Order SCANSORES.

Family RAMPHASTIDÆ.

Genus RAMPHASTOS.

51. *Ramphastos toco*, Gm.   Toco Toucan.
    Hab. Guiana.
    a. Presented by Viscount Powerscourt, F.Z.S., Aug. 24, 1860.

  b. Presented by H.E. Philip Edmund Wodehouse, C.B., Aug. 14, 1861.
  c. Deposited by George Dennis, Esq., Aug. 14, 1861.

52. *Ramphastos ariel,* Vig. Ariel Toucan.
 Hab. Brazil.

  a. Purchased, July 4, 1859.
  b. Received in exchange, Dec. 9, 1860.

53. *Ramphastos carinatus,* Swains. Sulphur-breasted Toucan.
 Hab. Mexico.

  a. Purchased, March 22, 1860.

Family MUSOPHAGIDÆ.

Genus CORYTHAIX.

54. *Corythaix buffonii* (Vieill.). Buffon's Touracou.
 Hab. West Africa.

  a. Presented by Admiral Sir H. Keppel, K.C.B., Aug. 14, 1861.
  b, c. Purchased, Dec. 15, 1860.

Genus MUSOPHAGA.

55. *Musophaga violacea,* Isert. Violaceous Plantain-cutter.
 Hab. West Africa.

  a. Purchased, Dec. 15, 1860.
  b. Deposited by Mr. Pinnock, Nov. 18, 1861.
  c, d. Deposited by Mr. Bartlett, March 21, 1861.

Family PSITTACIDÆ.

(Series NEOGEANA.)

Subfamily PSITTACULINÆ.

Genus PSITTACULA.

56. *Psittacula passerina* (Linn.). Passerine Parrakeet.
 Hab. Tropical America.

  a, b, c, d. Presented by — Davis, Esq., May 24, 1858.

PSITTACIDÆ.

Genus CAÏCA.

57. *Caïca melanocephala* (Linn.). Black-headed Parrot.
Hab. Demerara.
   a. Purchased, 1855.

Genus PIONUS.

58. *Pionus purpureus* (Gm.). Dusky Parrot.
Hab. Guiana.
   a. Purchased, Aug. 19, 1858.

Genus DEROPTYUS.

59. *Deroptyus accipitrinus* (Linn.). Hawk-headed Parrot.
Hab. Brazil.
   a. Presented by G. Dennis, Esq., July 19, 1856.

Genus CHRYSOTIS.

60. *Chrysotis pœcilorhynchus* (Shaw). Spotted-billed Amazon.
Hab. Venezuela.
   a. Purchased, May 21, 1860.

61. *Chrysotis levaillantii*, G. R. Gray. Levaillant's Amazon.
Hab. Mexico.
   a. Purchased, Aug. 3, 1859.

62. *Chrysotis ochrocephalus* (Gm.). Spanish-Main Amazon.
Hab. Trinidad.
   a. Purchased, July 16, 1860.

Subfamily ARINÆ.

Genus CONURUS.

63. *Conurus carolinensis* (Linn.). Carolinian Conure.
Hab. North America.
   a. Purchased, June 19, 1854.
   b. Purchased, Aug. 23, 1860.

64. *Conurus solstitialis* (Linn.). Yellow Conure.
Hab. Brazil.
   a. Purchased, April 25, 1862.

64*. *Conurus jandaya* (Gm.). Yellow-headed Conure.
Hab. Brazil.
  a. Purchased, June 17, 1854.

65. *Conurus monachus* (Bodd.). Grey-breasted Conure.
Hab. South America.
  a. Purchased, Sept. 14, 1860.
  b. Presented by Mrs. Malcolm, Aug. 5, 1859.

66. *Conurus aureus* (Linn.). Golden-crowned Conure.
Hab. South America.
  a, b. Purchased, April 27, 1849.

67. *Conurus acuticaudatus* (Vieill.). Sharp-tailed Conure.
Hab. South America.
  a. Purchased, May 20, 1853.

68. *Conurus erythrogenys* (Less.). Red-masked Conure.
Hab. Guayaquil.
  a. Purchased, 1854.

Genus ARA.

69. *Ara maracana* (Vieill.). Illiger's Maccaw.
Hab. Brazil.
  a. Purchased, June 7, 1861.

70. *Ara glauca*, Vieill. Glaucous Maccaw.
Hab. Brazil.
  a. Purchased, June 1860.

71. *Ara ararauna* (Linn.). Blue and Yellow Maccaw.
Hab. South America.
  a. Deposited by W. Duncan Stewart, Esq., May 24, 1861.
  b. Purchased, Aug. 24, 1859.

72. *Ara macao* (Linn.). Red and Blue Maccaw.
Hab. Central America.
  a. Deposited by H.E. W. D. Christie, F.Z.S., Minister to the Argentine Confederation, Nov. 7, 1859.

73. *Ara chloroptera*, G. R. Gray. Red and Yellow Maccaw.
Hab. South America.
  a. Deposited by W. Duncan Stewart, Esq., May 24, 1861.
  b. Presented by Mr. Atcheler, Jan. 2, 1862.

PSITTACIDÆ. 41

(Series PALÆOGEANA.)

Subfamily CACATUINÆ.

Genus LICMETIS.

74. *Licmetis tenuirostris* (Wagl.). Slender-billed Cockatoo.
 Hab. South Australia.
 a. Received in exchange from the Zoological Gardens, Amsterdam, May 17, 1847.

75. *Licmetis pastinator*, Gould. Western Slender-billed Cockatoo.
 Hab. Western Australia.
 a. Presented by Edgar Ray, Esq., Sept. 22, 1858.

Genus CACATUA.

76. *Cacatua citrino-cristata*, Fraser. Citron-crested Cockatoo.
 Hab. Timor Laut.
 a. Presented by Miss Julia Fox, Dec. 22, 1855.

77. *Cacatua leadbeateri* (Vig.). Leadbeater's Cockatoo.
 Hab. Australia.
 a. Presented by Lady Eleanor Cathcart, Nov. 21, 1854.
 b. Deposited by George Macleay, Esq., Aug. 19, 1859.
 c. Deposited by Samuel Gurney, Esq., June 1860.

78. *Cacatua triton* (Temm.). Triton Cockatoo.
 Hab. New Guinea.
 a. Purchased, 1860.

79. *Cacatua galerita* (Vieill.). Greater Sulphur-crested Cockatoo.
 Hab. Australia.
 a. Deposited by Mrs. Duff, 1860.
 b. Presented by Richard Tress, Esq., July 8, 1860.
 c. Deposited by Miss Lenox Conyngham, Dec. 15, 1860.
 d. Presented, 1854.

80. *Cacatua rosacea*, Vieill. Rose-crested Cockatoo.
 Hab. Moluccas.
 a. Deposited by Lady Caroline Duncombe, July 27, 1860.
 b. Deposited by Major Campbell, Roy. Art., Sept. 13, 1860.
 c. Deposited by Samuel Gurney, Esq., M.P., F.Z.S., Jan. 1861.
 d. Deposited by Mrs. Bainbridge, Aug. 30, 1855.
 e. Deposited by — Currey, Esq., F.Z.S., July 11, 1860.

81. *Cacatua sulphurea,* Vieill. Lesser Sulphur-crested Cockatoo.
   Hab. Moluccas.
   a. Presented by H. Wickens, Esq., Apr. 15, 1850.
   b. Presented by — Sutton, Esq., Oct. 23, 1855.
   c. Deposited by Miss Brown, 1860.

82. *Cacatua roseicapilla,* Vieill. Rosy Cockatoo.
   Hab. Australia.
   a. Presented by Lady Rolle, May 2, 1843.
   b. Deposited by the Countess Flahault, July 12, 1861.
   c. Deposited by George Macleay, Esq., C.M.Z.S., Aug. 19, 1859.

83. *Cacatua sanguinea,* Gould. Blood-stained Cockatoo.
   Hab. North Australia.
   a. Purchased, 1856.
   b. Purchased, 1858.

84. *Cacatua cristata* (Linn.). Greater White-crested Cockatoo.
   Hab. Moluccas.
   a. Purchased, July 19, 1861.

85. *Cacatua ducorpsii,* Hombr. & Jacq. Blue-eyed Cockatoo.
   Hab. Salomon Islands.
   a. Purchased, April 25, 1862.

Genus MICROGLOSSA.

86. *Microglossa alecto* (Temm.). Lesser Black Cockatoo.
   Hab. Aru Islands.
   a. Presented by Capt. Denham, R.N., June 20, 1861.

Genus CALLOCEPHALON.

87. *Callocephalon galeatum* (Lath.). Ganga Cockatoo.
   Hab. New South Wales.
   a. Male. Purchased, Aug. 5, 1859.

Genus CALYPTORHYNCHUS.

88. *Calyptorhynchus banksii.* Banksian Cockatoo.
   Hab. New South Wales.
   a. Purchased, April 25, 1862.

Subfamily LORIINÆ.

Genus TRICHOGLOSSUS.

89. *Trichoglossus hæmatodus* (Linn.). Blue-faced Lorikeet.
    Hab. Timor.
    a. Purchased, June 7, 1861.

Genus EOS.

90. *Eos riciniata*, Müll. et Schl. Blue-streaked Lory.
    Hab. Timor Laut.
    a. Purchased, Jan. 14, 1862.

Subfamily PSITTACINÆ.

Genus ECLECTUS.

91. *Eclectus cardinalis* (Bodd.). Linnean Eclectus.
    Hab. Moluccas.
    a. Presented by George Macleay, Esq., C.M.Z.S., Aug. 19, 1859.

Genus POLYCHLORUS.

92. *Polychlorus magnus* (Gm.). Red-sided Green-Parrot.
    Hab. Moluccas.
    a. Purchased, May 31, 1853.

Genus AGAPORNIS.

93. *Agapornis cana* (Gm.). Grey-headed Parrakeet.
    Hab. Madagascar.
    a, b. Purchased, March 17, 1860.
    c, d. Purchased, March 6, 1860.

94. *Agapornis pullaria* (Linn.). Love-bird Parrakeet.
    Hab. West Africa.
    a. Male. b. Female. Deposited by Mrs. Prehn, July 28, 1861.

Genus PŒOCEPHALUS.

95. *Pæocephalus meyerii* (Rüpp.). Meyer's Parrot.
    Hab. East Africa.
    a. Purchased, Jan. 18, 1855.

96. *Pæocephalus senegalensis* (Linn.).   Senegal Parrot.
   Hab. West Africa.
   a. Presented by Mrs. Clark, Oct. 11, 1853.

97. *Pæocephalus levaillantii* (Lath.).   Levaillant's Parrot.
   Hab. South Africa.
   a. Presented by Mrs. Jesse, June 14, 1853.

Genus PSITTACUS.

98. *Psittacus erythacus*, Linn.   Grey Parrot.
   Hab. West Africa.
   a. Purchased, 1854.

99. *Psittacus timneh*, Fraser.   Timneh Parrot.
   Hab. Sierra Leone.
   a. Purchased, Feb. 19, 1861.

Genus CORACOPSIS.

100. *Coracopsis vasa* (Linn.).   Greater Vasa Parrakeet.
   Hab. Madagascar.
   a. Presented by Mrs. David Barclay, June 12, 1827.
   b. Presented by Charles Telfair, Esq., C.M.Z.S., July 25, 1830.

101. *Coracopsis nigra* (Linn.).   Lesser Vasa Parrakeet.
   Hab. Madagascar.
   a. Purchased, Jan. 17, 1857.

Subfamily PLATYCERCINÆ.

Genus MELOPSITTACUS.

102. *Melopsittacus undulatus* (Shaw).   Undulated Grass Parrakeet.
   Hab. Australia.
   a. Bred in the Gardens, Oct. 30, 1859.
   b, c. Bred in the Gardens, 1860.
   d. Deposited by George Macleay, Esq., C.M.Z.S., June 22, 1859.

Genus EUPHEMA.

103. *Euphema pulchella* (Shaw).   Turquoisine Parrakeet.
   Hab. New South Wales.
   a. Male. b. Female. Presented by George Macleay, Esq., C.M.Z.S., June 22, 1859.

    c. Male. *d.* Female.   Bred in the Gardens, Aug. 10, 1861.
       From the above pair.
    e. Male. *f.* Female.   Bred in the Gardens, 1860.
    g. Bred in the Gardens, May 19, 1857.

104. *Euphema elegans*, Gould.   Elegant Grass Parrakeet.
    Hab. South Australia.
    a. Presented by Mr. Jamrach, Aug. 11, 1859.

Genus CALOPSITTA.

105. *Calopsitta novæ-hollandiæ* (Gm.).   Crested Ground Parrakeet.
    Hab. Australia.
    a. Presented by T. W. Nunn, Esq., March 13, 1862.
    b. Male. *c.* Female.   Received in exchange, April 17, 1862.

Genus PSEPHOTUS.

106. *Psephotus multicolor* (Brown).   Many-coloured Parrakeet.
    Hab. Australia.
    a. Female.   Purchased, March 9, 1861.
    b. Male. *c, d.* Females.   Purchased, April 25, 1862.

107. *Psephotus hæmatogaster*, Gould.   Blue-bonnet Parrakeet.
    Hab. Australia.
    a, b, c, d. Purchased, April 25, 1862.

108. *Psephotus hæmatonotus.*   Blood-rumped Parrakeet.
    Hab. Australia.
    a. Female.   Purchased, 1862.

Genus POLYTELIS.

109. *Polytelis barrabandii* (Swains.).   Barraband's Parrakeet.
    Hab. New South Wales.
    a. Male.   Presented by P. Fraser, Esq., May 30, 1847.

Genus PLATYCERCUS.

110. *Platycercus palliceps*, Vig.   Pale-headed Parrakeet.
    Hab. South Australia.
    a. Male.   Purchased, 1830.

111. *Platycercus pileatus*, Vig.   Pileated Parrakeet.
    Hab. Australia.
    a. Purchased, May 27, 1854.

46   PSITTACIDÆ.

112. *Platycercus eximius* (Shaw).   Rose-hill Parrakeet.
    Hab. New South Wales.
    a. Caught in the Gardens, 1856.
    b. Presented by G. H. Parkinson, Esq., Nov. 23, 1861.
    c. Purchased, Dec. 17, 1856.

113. *Platycercus pennantii* (Lath.).   Pennant's Parrakeet.
    Hab. New South Wales.
    a, b. Presented by Mrs. Wheeler, Aug. 20, 1861.
    c. Presented by R. P. Gunnell, Esq., June 21, 1853.
    d. Purchased, April 19, 1861.

114. *Platycercus adelaidæ*, Gould.   Adelaide Parrakeet.
    Hab. South Australia.
    a. Purchased, May 8, 1860.

115. *Platycercus barnardii* (Lath.).   Barnard's Parrakeet.
    Hab. South Australia.
    a, b. Purchased, April 25, 1853.

116. *Platycercus zonarius* (Shaw).   Bauer's Parrakeet.
    Hab. Australia.
    a, b. Purchased, March 9, 1861.

Genus APROSMICTUS.

117. *Aprosmictus erythropterus* (Lath.).   Red-winged Parrakeet.
    Hab. Australia.
    a. Female. Presented by R. Marshall, Esq., March 30, 1861.
    b. Male. Purchased, June 7, 1861.

118. *Aprosmictus scapulatus* (Bechst.).   King's Parrakeet.
    Hab. New South Wales.
    a. Male. Purchased, April 19, 1861.
    b. Female. Purchased, Dec. 9, 1859.

Subfamily PALÆORNITHINÆ.

Genus TANYGNATHUS.

119. *Tanygnathus macrorhynchus* (Linn.).   Great-billed Parrakeet.
    Hab. Gilolo and Ceram.
    a. Purchased, April 20, 1856.

PSITTACIDÆ. 47

120. *Tanygnathus müllerii* (Temm.). Müller's Great-billed Parrakeet.
Hab. Celebes.
a. Presented by the Babu Rajendra Mullick, July 14, 1857.

Genus PALÆORNIS.

121. *Palæornis javanica* (Osb.). Javan Parrakeet.
Hab. Java.
a. Purchased, 1853.
b. Purchased, 1857.
c. Purchased, Oct. 20, 1859.

122. *Palæornis luciani*, Verr. Red-cheeked Parrakeet.
Hab. East Indies.
a. Purchased, March 11, 1857.

123. *Palæornis bengalensis* (Linn.). Blossom-headed Parrakeet.
Hab. Hindostan.
a. Purchased, April 30, 1853.

124. *Palæornis columboïdes*, Vig. Malabar Parrakeet.
Hab. Southern India.
a. Purchased, June 2, 1852.

125. *Palæornis torquata* (Linn.). Ring-necked Parrakeet.
Hab. East Indies.
a. Male. Deposited by Richard Tress, Esq., March 8, 1861.
b. Female. Presented by George Dann, Esq., Sept. 13, 1861.

126. *Palæornis alexandri* (Linn.). Alexandrine Parrakeet.
Hab. Hindostan.
a. Female. Presented by Richard Tress, Esq., Oct. 22, 1855.

127. *Palæornis malaccensis* (Gm.). Malaccan Parrakeet.
Hab. Malacca.
a. Received in exchange, March 31, 1862.

Order FISSIROSTRES.

Family CAPRIMULGIDÆ.

Genus PODARGUS.

128. *Podargus cuvieri,* Vig. et Horsf. Cuvier's Podargus.
Hab. Van Diemen's Land.
   a. Purchased, Dec. 19, 1859.
   b. Purchased, Jan. 14, 1862.

Family ALCEDINIDÆ.

Genus ALCEDO.

129. *Alcedo ispida,* Linn. Kingfisher.
Hab. British Islands.
   a. Presented by J. C. Cumming, Esq., May 22, 1861.

Genus DACELO.

130. *Dacelo gigantea* (Lath.). Laughing Kingfisher.
Hab. Australia.
   a, b, c. Purchased, June 4, 1856.
   d. Presented by Dr. Müller, C.M.Z.S., May 11, 1860.
   e. Presented by Capt. Watson, July 4, 1861.

Family BUCEROTIDÆ.

Genus BUCEROS.

131. *Buceros ruficollis,* Temm. Red-necked Hornbill.
Hab. Salomon Islands.
   a. Male. Deposited by Dr. Bennett, F.Z.S., April 25, 1862.

Family UPUPIDÆ.

Genus UPUPA.

132. *Upupa epops,* Linn. Hoopoe.
Hab. British Islands.
   a. Purchased, July 1, 1861.

Family MOMOTIDÆ.

Genus MOMOTUS.

133. *Momotus subrufescens,* Sclater.   Carthagenian Motmot.
Hab. Carthagena.
   a. Purchased, July 17, 1860.

Order PASSERES.

Family ALAUDIDÆ.

Genus ALAUDA.

134. *Alauda arvensis,* Linn.   Skylark.
Hab. British Islands.
   a. Purchased, Nov. 3, 1860.

Family MOTACILLIDÆ.

Genus ANTHUS.

135. *Anthus pratensis* (Linn.).   Meadow Pipit.
Hab. British Islands.
   a. Purchased, Aug. 3, 1861.

Genus MOTACILLA.

136. *Motacilla yarrellii,* Gould.   Pied Wagtail.
Hab. British Islands.
   a, b. Purchased, Oct. 21, 1860.

Family SYLVIIDÆ.

Genus SIALIA.

137. *Sialia wilsoni,* Sw.   Common Blue-bird.
Hab. North America.
   a, b. Purchased, July 29, 1861.

Genus PHILOMELA.

138. *Philomela luscinia* (Linn.).   Nightingale.
Hab. British Islands.
   a. Presented by C. Clifton, Esq., F.Z.S., Oct. 1, 1861.

Genus SAXICOLA.

139. *Saxicola œnanthe,* Bechst.   Wheatear.
    Hab. British Islands.
    a. Purchased, May 14, 1861.

Family TURDIDÆ.

Genus TURDUS.

140. *Turdus boulboul,* Lath.   Varied-winged Thrush.
    Hab. Northern India.
    a. Presented, 1859.

141. *Turdus torquatus,* Linn.   Ring Ouzel.
    Hab. British Islands.
    a. Purchased, Nov. 15, 1860.

142. *Turdus viscivorus,* Linn.   Missel Thrush.
    Hab. British Islands.
    a. Presented by Mr. Travis, July 1856.

143. *Turdus migratorius,* Linn.   American Thrush.
    Hab. North America.
    a, b. Males.   c. Female.   Purchased, April 23, 1859.

144. *Turdus musicus,* Linn.   Song Thrush.
    Hab. British Islands.
    a. Purchased, 1861.

145. *Turdus merula,* Linn.   Blackbird.
    Hab. British Islands.
    a. Purchased, 1861.

146. *Turdus pilaris,* Linn.   Fieldfare.
    Hab. British Islands.
    a. Purchased, March 27, 1858.

Family PARIDÆ.

Genus SITTA.

147. *Sitta cæsia,* Meyer.   Common Nuthatch.
    Hab. British Islands.
    a. Purchased, April 30, 1860.

Family AMPELIDÆ.

Genus AMPELIS.

148. *Ampelis garrulus* (Linn.).   Waxwing Chatterer.
Hab. Europe.
   a, b. Presented by E. L. Preston, Esq., Oct. 2, 1861.

Family TANAGRIDÆ.

Genus PYRANGA.

149. *Pyranga rubra* (Linn.).   Scarlet Tanager.
Hab. North America.
   a, b. Males. Purchased, July 29, 1861.

Family FRINGILLIDÆ.

Subfamily PLOCEINÆ.

Genus VIDUA.

150. *Vidua paradisea* (Linn.).   Whydah Bird.
Hab. West Africa.
   a, b, c. Males. Purchased, Sept. 7, 1860.

151. *Vidua macroura* (Gm.).   Yellow-backed Whydah Bird.
Hab. West Africa.
   a. Purchased, Sept. 7, 1860.

Genus HYPHANTORNIS.

152. *Hyphantornis textor* (Gm.).   Rufous-necked Weaver Bird.
Hab. West Africa.
   a–e. Purchased, Aug. 24, 1859.

153. *Hyphantornis castaneo-fuscus*, Less.   Chestnut-backed Weaver Bird.
Hab. West Africa.
   a, b, c, d. Deposited by Mrs. Percy, Aug. 19, 1854.

Genus PLOCEUS.

154. *Ploceus philippinensis* (Linn.). Common Weaver Bird.
Hab. India.

a, b. Purchased, June 2, 1860.
c. Received in exchange, Nov. 7, 1853.
d. Deposited by Mrs. Sheldon, Aug. 10, 1858.

Genus EUPLECTES.

155. *Euplectes madagascariensis* (Linn.). Red-headed Weaver Bird.
Hab. Isle of France.

a. Purchased, April 21, 1858.

156. *Euplectes flammiceps,* Swains. Crimson-crowned Weaver Bird.
Hab. West Africa.

a. Purchased, April 21, 1858.

Subfamily ESTRELDINÆ.

Genus ESTRELDA.

157. *Estrelda ruficauda,* Gould. Red-tailed Finch.
Hab. New South Wales.

a, b, c. Purchased, April 28, 1861.

158. *Estrelda phaëton,* Homb. et Jacq. Crimson Finch.
Hab. Port Essington.

a, b, c, d. Males. *e, f, g.* Females. Presented by A. Denison, Esq., F.Z.S., June 5, 1861.

159. *Estrelda bichenovii* (Jard. et Selb.). Bicheno's Finch.
Hab. Queensland.

a, b, c. Presented by A. Denison, Esq., F.Z.S., June 5, 1861.

160. *Estrelda temporalis* (Lath.). Australian Waxbill.
Hab. Australia.

a, b, c, d. Purchased, Nov. 4, 1861.

161. *Estrelda cinerea* (Vieill.). Common Waxbill.
Hab. Western Africa.

a, b, c. Purchased, May 8, 1860.

Genus MUNIA.

162. *Munia oryzivora* (Linn.).   Java Sparrow.
   Hab. India.
   a, b. Purchased, Oct. 6, 1860.

163. *Munia leuconota* (Temm.), var.   Nutmeg Bird.
   Hab. Japan.
   a, b, c. Purchased, Oct. 6, 1860.

Genus POËPHILA.

164. *Poëphila cincta*, Gould.   Banded Grass-Finch.
   Hab. Queensland.
   a, b, c. Presented by A. Denison, Esq., F.Z.S., June 5, 1861.

Genus AMADINA.

165. *Amadina cucullata*, Swains.   Black-throated Finch.
   Hab. West Africa.
   a–g. Deposited by Mrs. Sheldon, Aug. 10, 1858.

166. *Amadina lathami* (Vig. et Horsf.).   Banded Grass-Finch.
   Hab. Australia.
   a, b, c, d. Presented by A. Denison, Esq., F.Z.S., June 5, 1861.

167. *Amadina modesta*, Gould.   Modest Grass-Finch.
   Hab. Australia.
   a. Purchased, May 7, 1862.

Genus DONACOLA.

168. *Donacola castaneothorax*, Gould.   Chestnut-breasted Finch.
   Hab. New South Wales.
   a–f. Presented by A. Denison, Esq., F.Z.S., June 5, 1861.

Subfamily EUSPIZINÆ.

Genus CYANOSPIZA.

169. *Cyanospiza cyanea* (Linn.).   Indigo Bird.
   Hab. North America.
   a. Male. Presented by G. Johnson, Esq., Aug. 30, 1860.
   b, c, d. Deposited by Mrs. Sheldon, Aug. 10, 1858.

Genus PAROARIA.

170. *Paroaria dominica* (Linn.). Red-headed Cardinal.
 Hab. Brazil.
 a, b. Presented by J. G. Leeming, Esq., Sept. 21, 1859.

Genus CARDINALIS.

171. *Cardinalis virginianus* (Briss.). Cardinal Grosbeak.
 Hab. North America.
 a, b, c. Purchased, April 24, 1856.

Genus GUBERNATRIX.

172. *Gubernatrix cristatellus* (Vieill.). Black-crested Cardinal.
 Hab. South America.
 a. Presented by J. G. Leeming, Sept. 21, 1859.

Genus ORYZOBORUS.

173. *Oryzoborus torridus* (Gm.). Tropical Seed-Finch.
 Hab. South America.
 a. Purchased, Aug. 8, 1860.

Genus TIARIS.

174. *Tiaris jacarini* (Linn.). Jacarini Finch.
 Hab. South America.
 a. Deposited by Mrs. Sheldon, Aug. 10, 1858.

Subfamily FRINGILLINÆ.

Genus LIGURINUS.

175. *Ligurinus chloris* (Linn.). Greenfinch.
 Hab. British Islands.
 a, b, c. Purchased, 1861.

Genus PASSER.

176. *Passer montanus* (Linn.). Tree Sparrow.
 Hab. British Islands.
 a, b, c. Purchased, Oct. 28, 1860.

FRINGILLIDÆ.

Genus FRINGILLA.

177. *Fringilla cœlebs,* Linn.   Chaffinch.
 Hab. British Islands.
 a, b, c. Purchased, 1861.

178. *Fringilla montifringilla,* Linn.   Mountain Finch.
 Hab. British Islands.
 a, b, c. Purchased, Dec. 3, 1859.

Genus ÆGIOTHUS.

179. *Ægiothus minor* (Ray).   Common Redpole.
 Hab. British Islands.
 a. Purchased, 1851.

180. *Ægiothus canescens* (Gould).   Mealy Redpole.
 Hab. British Islands.
 a. Purchased, Oct. 21, 1861.

181. *Ægiothus linarius* (Linn.).   Linnet.
 Hab. British Islands.
 a, b, c. Purchased, Dec. 26, 1860.

182. *Ægiothus montium* (Gm.).   Mountain Linnet or Twite.
 Hab. British Islands.
 a. Purchased, Dec. 26, 1860.

Genus CORYTHUS.

183. *Corythus enucleator* (Linn.).   Pine Grosbeak.
 Hab. British Islands.
 a. Purchased, July 29, 1861.

Subfamily EMBERIZINÆ.

Genus PLECTROPHANES.

184. *Plectrophanes nivalis* (Linn.).   Snow Bunting.
 Hab. Europe and N. Asia.
 a. Purchased, June 21, 1861.   From Japan.
 b. Purchased, June 30, 1853.

FRINGILLIDÆ.

Genus CRITHAGRA.

185. *Crithagra chrysopyga*, Swains.   Yellow-rumped Seed-Finch.
Hab. West Africa.
a, b. Purchased, Aug. 3, 1860.

Genus ZONOTRICHIA.

186. *Zonotrichia leucophrys* (Forst.).   White-eyebrowed Finch.
Hab. Labrador.
a. Presented by Lewis Henry Spence, Esq., Dec. 29, 1859.

Genus EMBERIZA.

187. *Emberiza hortulana* (Linn.).   Ortolan Bunting.
Hab. British Islands.
a, b, c, d. Purchased, Aug. 31, 1861.

188. *Emberiza cirlus* (Linn.).   Cirl Bunting.
Hab. British Islands.
a. Purchased, June 10, 1859.

189. *Emberiza* —— ? ——.   Japanese Bunting.
Hab. Japan.
a. Purchased, March 6, 1860.

190. *Emberiza citrinella* (Linn.).   Yellowhammer.
Hab. British Islands.
a, b. Purchased, 1861.

191. *Emberiza schœniclus*, Linn.   Black-headed Bunting.
Hab. British Islands.
a, b. Purchased, 1860.

192. *Emberiza miliaria*, Linn.   Common Bunting.
Hab. British Islands.
a, b. Purchased, 1861.

Genus CALAMOPHILUS.

193. *Calamophilus biarmicus* (Linn.).   Bearded Reedling.
Hab. British Islands.
a, c. Males. b, d. Females.   Purchased, Oct. 1, 1861.

Family STURNIDÆ.

Genus AGELÆUS.

194. *Agelæus phœniceus* (Linn.). Red-shouldered Starling.
Hab. North America.
a. Purchased, April 13, 1860.

Genus ICTERUS.

195. *Icterus jamaicai* (Gm.). Brazilian Hang-Nest.
Hab. Brazil.
a. Male. Presented by Frederick Bernal, Esq., H.B.M.'s Consul at Carthagena, New Granada, July 7, 1860.

Genus MOLOTHRUS.

196. *Molothrus badius* (Vieill.). Bay Troupial.
Hab. South Brazil.
a. Purchased, July 13, 1859.

Genus DOLICHONYX.

197. *Dolichonyx oryzivorus* (Linn.). Rice Bird.
Hab. North America.
a. Purchased, July 23, 1854.

Genus LAMPROCOLIUS.

198. *Lamprocolius chalybeus*, Ehrenb. Green Glossy Thrush.
Hab. West Africa.
a. Presented by Edward Cross, Esq., June 1, 1850.

199. *Lamprocolius auratus* (Gm.). Purple-headed Glossy Thrush.
Hab. West Africa.
a. Purchased, March 14, 1856.

Genus GRACULA.

200. *Gracula intermedia*, Hay. Indian Grakle.
Hab. Hindostan.
a. Presented by M. J. Harpley, Esq., Oct. 29, 1859.

Genus ACRIDOTHERES.

201. *Acridotheres cristatellus* (Linn.). Chinese Mynah.
 Hab. China.
 a, b. Purchased, Aug. 24, 1859.

202. *Acridotheres ginginianus* (Lath.). Indian Mynah.
 Hab. Hindostan.
 a. Purchased, June 30, 1860.

Genus STURNUS.

203. *Sturnus vulgaris,* Linn. Common Starling.
 Hab. British Islands.
 a, b. Caught in the Gardens, 1861.

Genus PTILONORHYNCHUS.

204. *Ptilonorhynchus holosericeus,* Kuhl. Bower Bird.
 Hab. New South Wales.
 a. Male. Presented by Thomas Walker, Esq., F.Z.S., Dec. 24, 1857.

Family PARADISEIDÆ.

Genus PARADISEA.

205. *Paradisea papuana,* Shaw. Lesser Bird of Paradise.
 Hab. New Guinea.
 a, b. Males. Purchased for the Society at Singapore and brought home by A. R. Wallace, Esq., F.Z.S., April 1, 1862.

Family CORVIDÆ.

Genus BARITA.

206. *Barita destructor,* Temm. Long-billed Butcher Bird.
 Hab. New Holland.
 a. Presented by John Dunn, Esq. March 30, 1860.

Genus GYMNORHINA.

207. *Gymnorhina leuconota,* Gould. White-backed Piping Crow.
 Hab. South Australia.
 a. Purchased, March 17, 1860.

208. *Gymnorhina organicum*, Gould. Tasmanian Piping Crow.
   Hab. Tasmania.
   a. Deposited by Mr. Macintosh, Feb. 24, 1859.
   b, c. Presented by Mrs. Hawker, Nov. 23, 1861.

Genus PICA.

209. *Pica caudata*, Flem. Magpie.
   Hab. British Islands.
   a. Presented by Miss Perry, Dec. 17, 1858.

Genus FREGILUS.

210. *Fregilus graculus*, Cuv. Cornish Chough.
   Hab. British Islands.
   a, b. Purchased, June 9, 1857.

Genus CYANOCITTA.

211. *Cyanocitta cristata* (Linn.). Blue Jay.
   Hab. North America.
   a, b. Purchased, 1859.
   c. Purchased, July 14, 1855.

Genus GARRULUS.

212. *Garrulus glandarius* (Linn.). Jay.
   Hab. British Islands.
   a. Purchased, Nov. 15, 1860.

Genus CORVUS.

213. *Corvus corone*, Linn. Carrion Crow.
   Hab. British Islands.
   a. Caught in the Gardens, 1861.

213*. *Corvus americanus*, Aud. American Crow.
   Hab. Nova Scotia.
   a, b. Presented by A. Downs, Esq., C.M.Z.S., April 2, 1862.

214. *Corvus monedula*, Linn. Jackdaw.
   Hab. British Islands.
   a, b. Caught in the Gardens, 1861.

215. *Corvus corax*, Linn. Raven.
   Hab. British Islands.
   a. Presented by Miss Prichard, Oct. 28, 1859.

216. *Corvus cornix*, Linn. Hooded Crow.
    Hab. British Islands.
    a. Purchased, Feb. 11, 1862.

Genus UROCISSA.

217. *Urocissa sinensis* (Bodd.). Chinese Magpie.
    Hab. China.
    a. Purchased, Feb. 19, 1861.

Order COLUMBÆ.

Family COLUMBIDÆ.

Genus TRERON.

218. *Treron chlorogaster*, Blyth. Fruit-eating Pigeon.
    Hab. India.
    a. Presented by H.R.H. Prince Dulcep Singh, June 24, 1861.

Genus CARPOPHAGA.

218*. *Carpophaga rubricera*, G. R. Gray. Red-eered Fruit-Pigeon.
    Hab. Salomon Islands.
    a. Deposited by Dr. George Bennett, F.Z.S., April 25, 1862.

Genus COLUMBA.

219. *Columba palumbus*, Linn. Ring Dove.
    Hab. British Islands.
    a. Presented by Charles Clifton, Esq., F.Z.S., Sept. 24, 1860.

220. *Columba œnas*, Linn. Stock Dove.
    Hab. British Islands.
    a. Purchased, 1860.

221. *Columba livia*, Linn. Rock Pigeon.
    Hab. British Islands.
    a. Male. b. Female. Presented by J. W. P. Orde, Esq., F.Z.S., Dec. 3, 1861.

222. *Columba leucocephala*, Linn. White-crowned Pigeon.
    Hab. West Indies.
    a. Presented by Thomas O'Connor Morris, Esq., Sept. 28, 1860.

223. *Columba maculosa*, Temm.   Grey Pigeon.
    Hab. South America.
    a, b. Presented by — Leigh, Esq., Feb. 20, 1851.

224. *Columba gymnophthalma*, Temm.   Naked-eyed Pigeon.
    Hab. South America.
    a. Presented by H.E. W. D. Christie, F.Z.S., Minister to the Argentine Confederation, Oct. 4, 1858.
    b. Hybrid between this species and *C. maculosa*, Temm.   Bred in the Gardens, Aug. 11, 1859.

225. *Columba speciosa*, Gm.   Necklaced Pigeon.
    Hab. South America.
    a. Purchased, July 26, 1860.

226. *Columba albilineata* (Bonap.).   White-naped Pigeon.
    Hab. South America.
    a. Purchased, May 19, 1853.

Genus ECTOPISTES.

227. *Ectopistes migratorius* (Linn.).   Passenger Pigeon.
    Hab. North America.
    a, b. Males.   Received in exchange, July 8, 1857.
    c. Male.   Presented by Earl Fitzwilliam, Feb. 11, 1858.
    d. Female.   Purchased, Aug. 3, 1860.
    e. Female.   Received in exchange, March 8, 1852.

Genus TURTUR.

228. *Turtur risorius* (Linn.).   Barbary Turtle Dove.
    Hab. Africa and India.
    a. Deposited by Robert Hudson, Esq., F.Z.S., Feb. 21, 1860.
    b, c. White variety.   Presented by a Lady, Dec. 7, 1860.

229. *Turtur senegalensis* (Linn.).   Cambayan Turtle Dove.
    Hab. Egypt.
    a, b. Bred in the Gardens, Sept. 1, 1861.
    c, d. Bred in the Gardens, Aug. 10, 1861.
    e, f, g, h. Purchased, 1860.
    i. Bred in the Gardens, Sept. 1, 1861.

230. *Turtur vinaceus* (Gm.).   Vinaceous Turtle Dove.
    Hab. West Africa.
    a. Presented by Earl Fitzwilliam, Feb. 11, 1858.

b. Bred in the Gardens, Oct. 18, 1858.
c. Bred in the Gardens, July 15, 1860.
d. Bred in the Gardens, Sept. 26, 1860.
e. Bred in the Gardens, June 27, 1861.

231. *Turtur malaccensis* (Lath.).   Barred Turtle Dove.
Hab. East Indies.

a. Bred in the Gardens, July 5, 1858.

Genus ZENAIDA.

232. *Zenaida amabilis,* Bp.   Zenaida Pigeon.
Hab. North America.

a, b, c. Purchased, 1861.

233. *Zenaida leucoptera* (Linn.).   White-winged Pigeon.
Hab. West Indies.

a. Purchased, May 30, 1860.

Genus CHAMÆPELIA.

234. *Chamæpelia passerina* (Linn.).   Passerine Ground Dove.
Hab. North America.

a, b, c. Presented by Sir Charles S. Smith, Sept. 29, 1860.

Genus PERISTERA.

235. *Peristera jamaicensis* (Linn.).   White-bellied Pigeon.
Hab. Jamaica.

a, b. Purchased, Oct. 28, 1861.

Genus GEOTRYGON.

236. *Geotrygon montana* (Linn.).   Red Ground Dove.
Hab. Brazil.

a, b. Purchased, Aug. 23, 1860.

237. *Geotrygon cristata* (Temm.).   Mountain Witch Dove.
Hab. Jamaica.

a, b. Purchased, July 30, 1861.
c. Purchased, Aug. 23, 1860.

Genus TYMPANISTRIA.

238. *Tympanistria bicolor,* Bp.   Tambourine Pigeon.
Hab. S. Africa.

a. Purchased, Oct. 1861.

COLUMBIDÆ.

Genus OCYPHAPS.

239. *Ocyphaps lophotes* (Temm.). Crested Pigeon.
   Hab. Australia.
   a. Bred in the Gardens, Aug. 21, 1859.

Genus CHALCOPHAPS.

240. *Chalcophaps indica* (Linn.). Green-winged Dove.
   Hab. India.
   a. Received in exchange. From China.

241. *Chalcophaps chrysochlora* (Wagl.). Little Green-winged Dove.
   Hab. Australia.
   a. Purchased, Feb. 18, 1861.

Genus PHAPS.

242. *Phaps chalcoptera* (Lath.). Bronze-winged Pigeon.
   Hab. Australia.
   a. Presented by George Macleay, Esq.,C.M.Z.S., June 22, 1859.
   b. Deposited by Charles Clifton, Esq., F.Z.S., Feb. 16, 1860.
   c. Bred in the Gardens, July 19, 1861.

Genus LEUCOSARCIA.

243. *Leucosarcia picata* (Lath.). Wonga-wonga Pigeon.
   Hab. New South Wales.
   a, b. Deposited by Richard Emery, Esq., March 5, 1856.
   c, d. Presented by George Macleay, Esq., C.M.Z.S., June 2, 1859.
   e. Bred in the Gardens, Aug. 11, 1859.

Genus PHLOGŒNAS.

244. *Phlogœnas cruentata* (Lath.). Red-breasted Pigeon.
   Hab. Philippine Islands.
   a. Received in exchange from the Société d'Acclimatation de Paris, July 15, 1861.

Subfamily GOURINÆ.

Genus GOURA.

245. *Goura victoriæ*, Fraser. The Victoria Crowned-Pigeon.
   Hab. Island of Jobie.
   a. Male. Purchased, Sept. 1, 1856.

b. Hybrid between this species and *G. coronata.* Bred in the Gardens, Aug. 5, 1850.
c, d. Purchased, Sept. 25, 1861.

246. *Goura coronata* (Linn.). The Common Crowned-Pigeon.
Hab. New Guinea.
   a. Male. Purchased, Sept. 5, 1852.

Order GALLINÆ.

Family CRACIDÆ.

Subfamily PENELOPINÆ.

Genus PENELOPE.

247. *Penelope purpurascens,* Wagl. Mexican Guan.
Hab. Central America.
   a, b. Presented by Chief Justice Temple, April 16, 1860. From Honduras.
   c. Presented by Robert Owen, Esq., C.M.Z.S., April 29, 1861. From Guatemala.

248. *Penelope superciliaris,* Wagl. White-eyebrowed Guan.
Hab. Brazil.
   a. Presented by H.G. the Duke of Richmond, April 24, 1862.

Genus ORTALIDA.

249. *Ortalida katraca* (Bodd.). Little Guan.
Hab. Trinidad.
   a, b. Presented by Dr. Huggins, C.M.Z.S., Sept. 16, 1861.

Subfamily CRACINÆ.

Genus CRAX.

250. *Crax carunculatus,* Temm. Yarrell's Curassow.
Hab. South America.
   a. Male. Purchased, April 5, 1859.
   b. Male. Purchased, July 19, 1854.
   c. Female. Purchased, March 12, 1861.

251. *Crax globicera*, Linn.   Globose Curassow.
   Hab. Honduras.
   a. Male. Presented by Mrs. Stevenson, April 27, 1857.
   b. Male. Purchased, 1861.
   c, d. Females. Presented by Chief Justice Temple, April 13, 1861.

252. *Crax albertii*, Fraser.   Prince Albert's Curassow.
   Hab. Brazil.
   a. Male. Purchased, Dec. 12, 1851.

253. *Crax alector*, Linn.   Crested Curassow.
   Hab. Guiana.
   a, b. Purchased, July 19, 1854.
   c. Presented by W. Duncan Stewart, Esq., June 26, 1861.

254. *Crax fasciolata*, Spix.   Banded Curassow.
   Hab. South America.
   a. Received in exchange, March 12, 1861.

255. *Crax blumenbachii*, Spix.   Blumenbach's Curassow.
   Hab. Brazil.
   a. Received in exchange, March 12, 1861.

Genus PAUXI.

256. *Pauxi mitu* (Linn.).   Razor-billed Curassow.
   Hab. Tropical America.
   a, b, c. Purchased, Aug. 6, 1860.

257. *Pauxi tomentosa* (Spix).   Lesser Razor-billed Curassow.
   Hab. Brazil.
   a, b. Purchased, Jan. 14, 1862.

Family PHASIANIDÆ.

Subfamily MELEAGRINÆ.

Genus MELEAGRIS.

258. *Meleagris ocellata*, Temm.   Ocellated Turkey.
   Hab. Guatemala.
   a. Female. Presented by Robert Owen, Esq., C.M.Z.S., April 29, 1861.

Subfamily PAVONINÆ.

Genus PAVO.

259. *Pavo cristatus*, Linn.   Common Pea-Fowl.
   Hab. India.

   a. Male. b. Female.   Deposited by Charles Clifton, Esq., F.Z.S., Jan. 16, 1860.
   c. Male. d. Female. (White variety.) Deposited by Charles Clifton, Esq., F.Z.S., Jan. 26, 1860.

260. *Pavo nigripennis*, Sclater.   Black-winged Pea-Fowl.
   Hab. India.

   a, b, c. Males. d, e. Females. Deposited by Charles Clifton, Esq., F.Z.S., Jan. 16, 1860.

261. *Pavo spicifer*, Vieill.   Green-necked Pea-Fowl.
   Hab. Malay Peninsula.

   a. Female.   Received in exchange from the Zoological Gardens, Rotterdam, May 2, 1860.

Genus POLYPLECTRON.

262. *Polyplectron chinquis*, Temm.   Peacock Pheasant.
   Hab. Northern India.

   a, b. Males.   Presented by the Babu Rajendra Mullick, July 14, 1857.

Subfamily PHASIANINÆ.

Genus THAUMALEA.

263. *Thaumalea picta* (Linn.).   Gold Pheasant.
   Hab. China.

   a, b. Males. c, d. Females.   Deposited by Mr. Bartlett, Nov. 1, 1860.
   e. Female.   Purchased, Feb. 2, 1860.

Genus CATREUS.

264. *Catreus wallichii* (Hardw.).   Cheer Pheasant.
   Hab. Northern India.

   a, b. Males.   Bred in the Gardens, July 12, 1859.
   c. Male.   Bred in the Gardens, July 4, 1860.
   d. Female.   Bred in the Gardens, July 17, 1860.
   e. Female.   Bred in the Gardens, June 27, 1861.

Genus PHASIANUS.

265. *Phasianus torquatus*, Gm.   Ring-necked Pheasant.
Hab. China.

a. Male. Presented by George Moss, Esq., May 23, 1859.
b. Male. Presented by Capt. Rees, July 21, 1860.
c, d. Females. Bred in the Gardens, July 27, 1861.

Genus NYCTHEMERUS.

266. *Nycthemerus argentatus*, Sw.   Silver Pheasant.
Hab. China.

a. Male. *b.* Female. Received in exchange, 1861.

Genus GALLOPHASIS.

267. *Gallophasis albo-cristatus* (Vig.).   White-crested Kaleege.
Hab. North-western Himalayas.

a. Male. *b.* Female. Presented by Viscount Canning, July 14, 1857.
c, d, e. Males. *f, g.* Females. Bred in the Gardens, 1859.
h, i. Females. Bred in the Gardens, July 9, 1860.
j–n. Females. Bred in the Gardens, July 9, 1861.

268. *Gallophasis melanotus* (Blyth).   Black-backed Kaleege.
Hab. Sikim.

a. Male. Presented by Viscount Canning, July 14, 1857.
b, c. Males. Bred in the Gardens, 1858.
d–g. Females. Presented by Viscount Canning, July 14, 1857.
h, i. Males. Bred in the Gardens, July 9, 1861.
j–o. Young. Bred in the Gardens, June 4, 1861.
p–s. Young. Bred in the Gardens, May 21, 1861.

269. *Gallophasis horsfieldii*, Gray.   Purple Kaleege.
Hab. North-western Himalayas.

a. Male. Presented by the Babu Rajendra Mullick, July 14, 1857.
b. Male. Bred in the Menagerie, 1858.
c, d, e, f. Males. Bred in the Menagerie, July 9, 1861.
g. Female. Presented by the Babu Rajendra Mullick, July 14, 1857.
h. Female. Bred in the Gardens, July 9, 1861.

Genus GALLUS.

270. *Gallus sonneratii*, Temm. Sonnerat's Jungle-Fowl.
    Hab. India.

    a, b. Males. Purchased, July 10, 1860.
    c–h. Females. Hybrids between the males of this species and the females of the Black-breasted Game-Bantam. Bred in the Gardens, 1861.

271. *Gallus furcatus*, Temm. Javan Jungle-Fowl.
    Hab. Java.

    a. Male. Hybrid between this species and Domestic Fowl. Purchased in Holland, 1861.

Subfamily LOPHOPHORINÆ.

Genus LOPHOPHORUS.

272. *Lophophorus impeyanus* (Lath.). Impeyan Pheasant.
    Hab. Himalaya Mountains.

    a, b. Males. c. Female. Bred in the Gardens, 1853.
    d, e, f. Females. Bred in the Gardens, June 10, 1861.
    g, h, i. Females. Bred in the Gardens, June 20, 1861.

Family TETRAONIDÆ.

Subfamily PERDICINÆ.

Genus PTILOPACHUS.

273. *Ptilopachus fuscus* (Vieill.). Buff-breasted Partridge.
    Hab. West Africa.

    a, b. Purchased, Jan. 4, 1862.

Genus FRANCOLINUS.

274. *Francolinus capensis*, Gm. Cape Francolin.
    Hab. South Africa.

    a, b, c. Presented by H.E. Sir George Grey, K.C.B., F.Z.S., Governor of New Zealand, Oct. 3, 1859.
    d. Bred in the Gardens, Aug. 4, 1861.

275. *Francolinus ponticerianus* (Gm.). Grey Francolin.
    Hab. India.

    a–g. Purchased, April 5, 1862.

Genus COTURNIX.

276. *Coturnix coromandelica* (Gm.). Rain Quail.
   Hab. India.

   a, b, c. Presented by Prince Duleep Singh, June 24, 1861.

Genus CACCABIS.

277. *Caccabis rufa* (Linn.). Barbary Partridge.
   Hab. North Africa.

   a. Received in exchange, April 16, 1861.
   b. Presented by Col. the Hon. Hely Hutchinson, F.Z.S., May 3, 1862.

Genus SYNŒCUS.

278. *Synœcus australis* (Lath.). Australian Quail.
   Hab. Australia.

   a, b, c. Presented by Dr. Müller, C.M.Z.S., Feb. 27, 1861.

Subfamily ODONTOPHORINÆ.

Genus ORTYX.

279. *Ortyx virginianus* (Linn.). Virginian Colin.
   Hab. North America.

   a–h. Presented by A. Downs, Esq., of Halifax, C.M.Z.S., March 5, 1861.

280. *Ortyx cubanensis*, Gould. Cuban Colin.
   Hab. Cuba.

   a. Presented by Lord Harris, F.Z.S., July 7, 1851.

Genus LOPHORTYX.

281. *Lophortyx californianus* (Lath.). Californian Quail.
   Hab. California.

   a, b. Males. c. Female. Deposited by Mr. Bartlett, Sept. 10, 1859.
   d, e, f, g. Bred in the Gardens, July 5, 1860.

Genus DENDRORTYX.

282. *Dendrortyx* ——? ——. Guatemalan Tree-Partridge.
   Hab. Guatemala.

   a. Presented by Robert Owen, Esq., C.M.Z.S., April 29, 1861.

Subfamily TETRAONINÆ.

Genus TETRAO.

283. *Tetrao cupido*, Linn. Prairie Grouse.
 Hab. North America.
   a–k. Deposited by J. Stone, Esq., May 3, 1862.

Genus BONASA.

284. *Bonasa umbellus* (Linn.). Ruffed Grouse.
 Hab. Nova Scotia.
   a, b, c. Presented by A. Downs, Esq., of Halifax, C.M.Z.S., Dec. 23, 1861.

Subfamily PTEROCLINÆ.

Genus PTEROCLES.

285. *Pterocles alchata* (Linn.). Pintailed Sand Grouse.
 Hab. Southern Europe.
   a. Male. b. Female. Purchased, Oct. 7, 1861.
   c. Male. Deposited by Mr. Bartlett, 1861.

Genus SYRRHAPTES.

286. *Syrrhaptes paradoxus* (Pall.). Pallas's Sand Grouse.
 Hab. China.
   a–o. Presented by the Hon. J. F. Stuart Wortley, April 15, 1861.
   p, q. Presented by Capt. Hand, R.N., April 29, 1861.
   r. Presented by Capt. Commerell, R.N., V.C., May 9, 1861.
   s, t, u. Presented by A. O'Brien, Esq., Feb. 27, 1862.

Family MEGAPODIIDÆ.

Genus TALEGALLA.

287. *Talegalla lathami*, Gray. Brush-Turkey.
 Hab. Australia.
   a. Female. Deposited by the Duke of Buccleuch, 1856.
   b. Female. Bred in the Gardens, July 19, 1854.
   c. Female. Bred in the Gardens, Aug. 29, 1860.

Family TINAMIDÆ.

Genus TINAMUS.

288. *Tinamus variegatus*, Gm.   Variegated Tinamou.
Hab. Brazil.

    a. Presented by John Blount, Esq., April 8, 1861.

Order STRUTHIONES.

Family STRUTHIONIDÆ.

Genus STRUTHIO.

289. *Struthio camelus,* Linn.   Ostrich.
Hab. North Africa.

    a. Male. *b.* Female. Presented by Her Majesty The Queen, Feb. 8, 1859. From Morocco.
    c. Male. *d.* Female. Var. *meridionalis.* From South Africa. Presented by H.E. Sir George Grey, K.C.B., F.Z.S., Governor of New Zealand, Nov. 1, 1861.

Genus RHEA.

290. *Rhea darwinii,* Gould.   Darwin's Rhea.
Hab. South America.

    a. Male.   Purchased, Oct. 20, 1858.

291. *Rhea macrorhyncha,* Sclater.   Great-billed Rhea.
Hab. South America.

    a. Male.   Purchased, Nov. 1858.

292. *Rhea americana,* Vieill.   Common Rhea.
Hab. South America.

    a. Female. Presented by George Wilks, Esq., July 15, 1856.
    b. Male. *c.* Female. Purchased, Dec. 5, 1860.

Genus CASUARIUS.

293. *Casuarius galeatus* (Vieill.).   Common Cassowary.
Hab. Ceram.

    a Received in exchange from the Zoological Gardens, Rotterdam, March 28, 1862.

293*. *Casuarius bennettii,* Gould.   Mooruk.
   Hab. New Britain.
   a. Male. Presented by Dr. George Bennett, F.Z.S., May 17, 1857.
   b. Male. c. Female. Presented by Dr. George Bennett, F.Z.S., May 25, 1858.

Genus DROMÆUS.

294. *Dromæus novæ-hollandiæ,* Vieill.   Emeu.
   Hab. New South Wales.
   a. Male. Presented by the Marchioness of Londonderry, Feb. 14, 1857.
   b. Female. Received in exchange from Viscount Hill, F.Z.S., 1856.

295. *Dromæus irroratus,* Bartlett.   Spotted Emeu.
   Hab. Western Australia.
   a, b. Purchased, May 18, 1860.
   c, d. Deposited by — Bennett, Esq. Bred in the Gardens, 1861.

Family APTERYGIDÆ.

Genus APTERYX.

296. *Apteryx mantellii,* Bartlett.   Kiwi.
   Hab. New Zealand.
   a. Female. Presented by Lieut. Governor Eyre, Dec. 9, 1851.

Order GRALLÆ.

Family OTIDÆ.

Genus TETRAX.

297. *Tetrax campestris,* Leach.   Little Bustard.
   Hab. Europe.
   a. Received in exchange, Sept. 20, 1861.

Genus EUPODOTIS.

298. *Eupodotis bengalensis* (Gm.).   Bengal Bustard.
   Hab. Bengal.
   a. Presented by the Babu Rajendra Mullick, July 14, 1857.

Family CHARADRIIDÆ.

Genus VANELLUS.

299. *Vanellus cristatus*, Meyer. Peewit.
 Hab. British Islands.
 a. Deposited by John Scott, Esq., F.Z.S., 1861.

Genus HOPLOPTERUS.

300. *Hoplopterus tectus* (Bodd.). Hooded Plover.
 Hab. West Africa.
 a. Purchased, Oct. 19, 1851.

Genus ŒDICNEMUS.

301. *Œdicnemus crepitans*, Temm. Thicknee.
 Hab. British Islands.
 a, b, c, d. Presented by Lord Lilford, F.Z.S., June 17, 1861.

Family TOTANIDÆ.

Genus HÆMATOPUS.

302. *Hæmatopus ostralegus*, Linn. Oyster-catcher.
 Hab. Europe, Asia, Africa.
 a. Purchased, 1858.

Family SCOLOPACIDÆ.

Genus NUMENIUS.

303. *Numenius phæopus*, Linn. Common Whimbrel.
 Hab. Europe, Asia, North Africa.
 a. Presented by Mrs. Statham, Sept. 15, 1855.
 b. Purchased, July 6, 1860.

304. *Numenius arquatus*, Linn. Common Curlew.
 Hab. Europe, Asia, Africa.
 a. Purchased, March 6, 1860.

Genus LIMOSA.

305. *Limosa lapponica*, Linn. Bar-tailed Godwit.
 Hab. Europe.
 a, b. Deposited by Lord Lilford, F.Z.S., Aug. 11, 1860.

306. *Limosa melanura*, Leisler. Black-tailed Godwit.
 Hab. Europe.
 a. Purchased, Aug. 10, 1860.

Genus MACHETES.

307. *Machetes pugnax* (Linn.). Ruff.
  Hab. Europe.
  a–e. Purchased, July 20, 1860.

Family PSOPHIIDÆ.

Genus PSOPHIA.

308. *Psophia crepitans*, Linn. Common Trumpeter.
  Hab. Guiana.
  a. Presented by Sir William H. Holmes, July 23, 1861.

Genus CARIAMA.

309. *Cariama cristata*, Linn. Cariama.
  Hab. Tropical America.
  a. Purchased, Feb. 25, 1851.

Family GRUIDÆ.

Genus GRUS.

310. *Grus montignesia*, Bonap. Mantchourian Crane.
  Hab. North China.
  a, b. Males. Presented by Her Majesty The Queen, Feb. 25, 1857.
  c. Female. Received in exchange from the Zoological Gardens, Paris, Oct. 20, 1856.
  d. Young. Bred in the Gardens, June 24, 1861.

311. *Grus antigone*, Linn. Sarus Crane.
  Hab. Northern India.
  a. Presented by the Babu Rajendra Mullick, July 14, 1857.

312. *Grus australasiana*, Gould. Australian Crane.
  Hab. Australia.
  a. Male. Presented by the Marchioness of Londonderry, Feb. 14, 1857.
  b. Female. Purchased, Dec. 19, 1859.

313. *Grus cinerea*, Bechst. Common Crane.
  Hab. Europe, Asia, Africa.
  a. Male. Received in exchange from the Zoological Gardens, Amsterdam, May 27, 1852.
  b. Female. Purchased, May 13, 1848.

314. *Grus carunculata* (Gm.).   Wattled Crane.
Hab. South Africa.

  a. Presented by H.E. Sir George Grey, K.C.B., F.Z.S., Governor of New Zealand, May 26, 1861.

Genus TETRAPTERYX.

315. *Tetrapteryx paradiseus*, Licht.   Stanley Crane.
Hab. South Africa.

  a, b, c. Presented by H.E. Sir George Grey, K.C.B., F.Z.S., Governor of New Zealand, Nov. 1, 1861.
  d. Presented by H.E. Sir George Grey, K.C.B., F.Z.S., Governor of New Zealand, May 26, 1861.

Genus BALEARICA.

316. *Balearica pavonina*, Briss.   Balearic Crowned Crane.
Hab. North and West Africa.

  a, b. Presented by Earl Fitzwilliam, Feb. 11, 1858.
  c. Presented by Viscount Hill, F.Z.S., Feb. 29, 1860.

317. *Balearica regulorum*, Licht.   Cape Crowned Crane.
Hab. South Africa.

  a. Deposited by H.E. Sir George Grey, K.C.B., F.Z.S., Governor of New Zealand, Feb. 19, 1851.

Family ARDEIDÆ.

Subfamily ARDEINÆ.

Genus ARDEA.

318. *Ardea cinerea*, Linn.   Common Heron.
Hab. Europe.

  a. Presented by the Hon. C. A. Ellis, F.Z.S., Oct. 15, 1859.
  b. Presented by Lord Lilford, Aug. 1, 1855.

319. *Ardea purpurea*, Linn.   Purple Heron.
Hab. Europe.

  a, b, c, d. Received in exchange from the Zoological Gardens, Cologne, Oct. 21, 1861.

Genus ARDETTA.

320. *Ardetta minuta* (Linn.).   Little Bittern.
Hab. Newfoundland; Europe.

  a, b. Purchased, June 8, 1860.

Genus HERODIAS.

321. *Herodias russata,* Temm. Buff-backed Heron.
 Hab. Egypt.
 a. Purchased, April 21, 1853.

Genus TIGRISOMA.

322. *Tigrisoma tigrinum* (Gm.). Tiger Bittern.
 Hab. Tropical America.
 a. Presented by Capt. Moss, July 7, 1851.

323. *Tigrisoma leucolophum,* Jard. White-crested Tiger Bittern.
 Hab. West Africa.
 a. Purchased, April 2, 1862.

Genus EURYPYGA.

324. *Eurypyga helias* (Pall.). Sun Bittern.
 Hab. Trinidad.
 a. Presented by Dr. Huggins, C.M.Z.S., Sept. 16, 1861.

Genus RHINOCHETUS.

325. *Rhinochetus jubatus,* Verr. et Des Murs. Kagu.
 Hab. New Caledonia.
 a. Presented by Dr. G. Bennett, F.Z.S., April 1862.

Subfamily PLATALEINÆ.

Genus PLATALEA.

326. *Platalea leucorodia,* Linn. Spoonbill.
 Hab. Europe, Asia, Africa.
 a, b. Received in exchange from the Zoological Gardens, Amsterdam, March 30, 1860.

Subfamily CICONIINÆ.

Genus CICONIA.

327. *Ciconia nigra,* Ray. Black Stork.
 Hab. Europe.
 a. Presented by W. C. Domville, Esq., F.Z.S., Aug. 14, 1855.
 b, c. Received in exchange from the Zoological Gardens, Cologne, Oct. 21, 1861.

328. *Ciconia alba*, Briss.   White Stork.
   Hab. Europe.
   a. Presented by W. C. Domville, Esq., F.Z.S., Aug. 14, 1855.

329. *Ciconia maguari*, Briss.   Maguari Stork.
   Hab. South America.
   a. Presented by Mr. Jamrach, July 1, 1847.

Genus LEPTOPTILUS.

330. *Leptoptilus crumeniferus* (Cuv.).   Marabou Stork.
   Hab. West Africa.
   a. Presented by Edmund Gabriel, Esq., H.B.M.'s Commissioner at Loando, Angola, Aug. 22, 1860. From Angola.
   b. Purchased, Aug. 15, 1854.

Genus MYCTERIA.

331. *Mycteria senegalensis* (Shaw).   Saddle-billed Jabiru.
   Hab. West Africa.
   a, b. Purchased, April 17, 1861.

Family TANTALIDÆ.

Genus IBIS.

332. *Ibis rubra*, Linn.   Scarlet Ibis.
   Hab. Demerara.
   a. Purchased, 1857.

Genus GERONTICUS.

333. *Geronticus æthiopicus* (Lath.).   Sacred Ibis.
   Hab. River Gambia.
   a, b. Received in exchange from the Zoological Gardens, Antwerp, Aug. 9, 1855.
   c. Purchased, Nov. 8, 1851.

334. *Geronticus calvus* (Bodd.).   Bald-headed Ibis.
   Hab. South Africa.
   a. Presented by H.E. Sir George Grey, K.C.B., F.Z.S., Governor of New Zealand, Oct. 3, 1851.

Family RALLIDÆ.

Genus RALLUS.

335. *Rallus pectoralis*, Less.   Australian Rail.
 Hab. New Holland.

 a, b. Received in exchange, May 15, 1860.

Genus CREX.

336. *Crex pratensis*, Bechst.   Land Rail.
 Hab. British Islands.

 a. Purchased, Sept. 12, 1861.

Genus ARAMIDES.

337. *Aramides cayennensis* (Gm.).   West-Indian Rail.
 Hab. Trinidad.

 a. Presented by Dr. Huggins of Trinidad, C.M.Z.S., Sept. 16, 1861.

Genus OCYDROMUS.

338. *Ocydromus australis* (Sparrm.).   Weka Rail.
 Hab. New Zealand.

 a. Male. *b.* Female. Purchased, April 29, 1861.
 c. Male. Purchased, Jan. 21, 1859.

Genus PORPHYRIO.

339. *Porphyrio smaragnotis*, Temm.   Egyptian Porphyrio.
 Hab. North Africa.

 a. Presented by Earl Fitzwilliam, Feb. 11, 1858.

340. *Porphyrio melanotus*, Temm.   Black-backed Porphyrio.
 Hab. Australia.

 a. Presented by Edward Wilson, Esq., May 31, 1860.

341. *Porphyrio* —— ? ——.   Pacific Porphyrio.
 Hab. Island of Tanna, New Hebrides.

 a. Deposited by Dr. G. Bennett, F.Z.S., April 1862.

Genus GALLINULA.

342. *Gallinula chloropus* (Linn.).   Common Waterhen.
 Hab. British Islands.

 a. Purchased, June 1861.

343. *Gallinula nesiotis*, Sclater.   Island Hen.
 Hab. Tristan d'Acunha.

  a. Presented by H.E. Sir George Grey, K.C.B., F.Z.S., Governor of New Zealand, May 26, 1861.

Genus FULICA.

344. *Fulica cristata*, Lath.   Crested Coot.
 Hab. South Africa.

  a. Presented by Earl Fitzwilliam, Feb. 11, 1858.

345. *Fulica atra*, Linn.   Common Coot.
 Hab. Europe.

  a. Purchased, Nov. 15, 1860.
  b. Purchased, Jan. 3, 1859.

Order ANSERES.

Family ANATIDÆ.

Subfamily ANSERINÆ.

Genus ANSERANAS.

346. *Anseranas melanoleuca*, Less.   Black and White Goose.
 Hab. Australia.

  a. Purchased, June 19, 1855.
  b. Presented by Dr. F. Müller of Melbourne, C.M.Z.S., May 20, 1862.

Genus PLECTROPTERUS.

347. *Plectropterus gambensis* (Linn.).   Spur-winged Goose.
 Hab. West Africa.

  a. Purchased, June 25, 1857.

Genus CEREOPSIS.

348. *Cereopsis novæ-hollandiæ*, Lath.   Cereopsis Goose.
 Hab. Australia.

  a. Male.   Bred in the Gardens, 1853.
  b, c. Females.   Purchased, April 26, 1861.

Genus ANSER.

349. *Anser ferus*, Linn.   Wild or Grey-lag Goose.
 Hab. British Islands, Europe, Asia.

  a. Purchased, 1855.   From India.
  b. Purchased, March 24, 1860.

80   ANATIDÆ.

350. *Anser brachyrhynchus*, Baill.   Pink-footed Goose.
    Hab. British Islands.
    a, b. Purchased, March 14, 1861.

351. *Anser segetum*, Linn.   Bean Goose.
    Hab. Europe.
    a, b. Purchased, Nov. 1, 1859.

352. *Anser erythropus*, Linn.   Little Goose.
    Hab. Europe.
    a. Male.   Purchased, April 4, 1858.
    b. Female. Purchased, May 13, 1852.
    c. Male.   Purchased, May 27, 1852.

353. *Anser albifrons*.   White-fronted Goose.
    Hab. Europe.
    a, b. Purchased, 1861.

354. *Anser indicus*, Gm.   Bar-headed Goose.
    Hab. India.
    a. Male. b. Female. Received in exchange, Feb. 5, 1852.

355. *Anser cygnoïdes*, Linn.   Chinese Goose.
    Hab. China.
    a. Male. Presented by Capt. Cruikshank, April 22, 1860.
    b. Female. Presented by Russell Sturgis, Esq., F.Z.S., Oct. 7, 1859.

Genus BERNICLA.

356. *Bernicla canadensis* (Linn.).   Canada Goose.
    Hab. British Islands.
    a. Male. b. Female. Presented by Capt. Wishart, Oct. 29, 1861.
    c. Male. d. Female. Hybrid between this species and *Anser ferus*, Linn. Deposited by Mr. Bartlett, Sept. 10, 1859.

357. *Bernicla hutchinsii*, Rich.   Hutchins's Goose.
    Hab. Fort Churchill, Hudson's Bay Territory.
    a. Female. Presented by Capt. Herd, Oct. 10, 1860.

358. *Bernicla brenta* (Flem.).   Brent Goose.
    Hab. Europe.
    a. Male. b. Female. Purchased, 1860.

359. *Bernicla leucopsis*, Bechst.   Bernicle Goose.
   Hab. Europe.
   a. Purchased, Nov. 1, 1860.

360. *Bernicla ruficollis* (Pall.).   Red-breasted Goose.
   Hab. Europe.
   a. Female. Received in exchange from the Zoological Gardens, Amsterdam, Aug. 19, 1858.

Genus CHLOËPHAGA.

361. *Chloëphaga magellanica* (Gm.).   Upland Goose.
   Hab. Falkland Islands.
   a. Male. b. Female. Presented by H.E. Capt. Moore, R.N., Governor of the Falkland Islands, May 27, 1857.
   c. Female. Purchased, July 5, 1860.
   d. Male. e. Female. Presented by H.E. Capt. Moore, R.N., Governor of the Falkland Islands, Sept. 5, 1861.
   f. Young. Bred in the Gardens, April 27, 1862.

362. *Chloëphaga poliocephala*, G. R. Gray.   Ashy-headed Goose.
   Hab. S. America.
   a. Male. b. Female. Bred in the Gardens, June 7, 1858.
   c. Male. d. Female. Bred in the Gardens, May 27, 1860.

363. *Chloëphaga rubidiceps*, Sclater.   Ruddy-headed Goose.
   Hab. Falkland Islands.
   a, b. Males. c, d. Females. Purchased, July 5, 1860.

364. *Chloëphaga sandvicensis* (Vig.).   Sandwich-Island Goose.
   Hab. Sandwich Islands.
   a. Male. Presented by Earl Fitzwilliam, Feb. 11, 1858.
   b. Female. Presented by Viscount Hill, F.Z.S., March 22, 1842.
   c. Female. Received in exchange, Aug. 3, 1847.

Subfamily CYGNINÆ.

Genus CYGNUS.

365. *Cygnus olor* (Gm.).   Common Swan.
   Hab. British Islands.
   a. Presented by the Duke of Northumberland, 1861.

366. *Cygnus nigricollis* (Gm.).   Black-necked Swan.
   Hab. Chili.
   a. Male. Purchased, Oct. 19, 1851.

b. Female. Presented by Capt. the Hon. E. A. Harris, R.N., H.M.'s Commissioner-General in Chili, C.M.Z.S., Dec. 3, 1856.
c. Male. Bred in the Gardens, June 23, 1858.
d. Female. Bred in the Gardens, June 27, 1859.

367. *Cygnus atratus,* Lath. Black Swan.
Hab. Australia.
a. Male. *b.* Female. Purchased, Jan. 15, 1858.
c. Male. Presented by Dr. Müller, C.M.Z.S., Sept. 19, 1860.

Subfamily ANATINÆ.

Genus DENDROCYGNA.

368. *Dendrocygna autumnalis* (Linn.). Red-billed Tree-Duck.
Hab. Tropical America.
a, b. Presented by W. Duncan Stewart, Esq., F.Z.S., June 26, 1861.
c. Presented by Mrs. S. C. Hall, Nov. 17, 1855.
d. Purchased, Aug. 8, 1857.

369. *Dendrocygna viduata* (Linn.). White-faced Tree-Duck.
Hab. Brazil.
a, b, c. Presented by H.E. W. D. Christie, F.Z.S., Minister to Brazil, May 28, 1862.

370. *Dendrocygna arcuata* (Cuv.). Indian Tree-Duck.
Hab. India.
a, b, c. Purchased, Sept. 3, 1858.
d. Presented by the Babu Rajendra Mullick, July 14, 1857.

Genus TADORNA.

371. *Tadorna vulpanser,* Flem. Common Sheldrake.
Hab. Europe, Asia, America.
a. Male. *b.* Female. Purchased, June 24, 1860.
c. Male. *d.* Female. Hybrids between this species and *Casarca cana* (Gm.). Bred in the Gardens, June 26, 1859.

Genus CASARCA.

372. *Casarca rutila* (Pall.). Ruddy Sheldrake.
Hab. Europe, Asia, Africa.
a. Male. *b.* Female Bred in the Gardens, May 13, 1859.
c. Male. *d.* Female. Purchased, May 28, 1850. From Egypt.

ANATIDÆ. 83

   e. Male. *f.* Female. Bred in the Gardens, June 2, 1861.
   g. Male. Hybrid between Indian specimen of this species and female *Casarca cana* (Gm.).

373. *Casarca cana** (Gmel.). White-faced Sheldrake.
   Hab. South Africa.
   a. Female. Purchased, Nov. 8, 1851.

374. *Casarca tadornoïdes*, Jard. et Selb. Australian Sheldrake.
   Hab. South Australia.
   a–d. Females. Presented by the Hon. J. C. Hawker, Speaker of the House of Assembly, Adelaide, April 3, 1862.

Genus AIX.

375. *Aix sponsa* (Linn.). Summer Duck.
   Hab. North America.
   a. Male. *b, c.* Females. Bred in the Gardens, May 24, 1859.
   d. Male. Hybrid between this species and *Fuligula ferina* (Linn.). Received in exchange from the Zoological Gardens, Amsterdam, Nov. 11, 1860.
   e, f. Males. Hybrids between this species and *Nyroca leucophthalma* (Bechst.). Received in exchange from the Zoological Gardens, Amsterdam, Nov. 11, 1860.

376. *Aix galericulata* (Linn.). Mandarin Duck.
   Hab. China.
   a, b. Males. *c.* Female. Bred in the Gardens, June 2, 1859.
   d. Female. Received in exchange from the Zoological Gardens, Antwerp, Dec. 23, 1858.
   e, f. Females. Purchased, Sept. 20, 1860.

Genus MARECA.

377. *Mareca penelope* (Linn.). Wigeon.
   Hab. Europe.
   a. Presented by Earl Fitzwilliam, Feb. 11, 1858.
   b. Male. *c.* Female. Supposed hybrids between this species and the Teal (*Querquedula crecca*).

Genus DAFILA.

378. *Dafila acuta* (Linn.). Pintail.
   Hab. Europe.
   a, b. Males. *c.* Female. Bred in the Gardens, May 15, 1861.
   d, e, f. Females. Bred in the Gardens, May 16, 1860.

* For hybrids from this specimen see *Casarca rutila* and *Tadorna vulpanser*.

Genus Pœcilonetta.

379. *Pœcilonetta bahamensis* (Linn.). Bahama Duck.
   Hab. West Indies.
   a–g. Bred in the Gardens, June 4, 1861.
   h. Received in exchange from the Zoological Gardens, Rotterdam, Dec. 3, 1860.
   i, j, k. Bred in the Gardens, July 31, 1860.

380. *Pœcilonetta erythrorhyncha*, Gm. Red-billed Duck.
   Hab. South Africa.
   a. Male. b. Female. Bred in the Gardens, July 9, 1860.
   c. Male. d. Female. Bred in the Gardens, July 5, 1859.

Genus Anas.

381. *Anas obscura*, Gm. Dusky Duck.
   Hab. North America.
   a. Male. b. Female. Bred in the Gardens, May 21, 1861.

382. *Anas xanthorhyncha*, Forst. Yellow-billed Duck.
   Hab. South Africa.
   a. Male. b. Female. Bred in the Gardens, May 20, 1859.
   c. Male. d. Female. Bred in the Gardens, May 30, 1860.
   e. Male. Hybrid between this species and *Anas boschas*, Linn.

383. *Anas superciliosa*, Gm. Australian Wild Duck.
   Hab. Australia.
   a. Female. Presented by Edward Wilson, Esq., of Melbourne, May 31, 1860.

Genus Querquedula.

384. *Querquedula crecca* (Linn.). Common Teal.
   Hab. Europe.
   a. Male. b. Female. Bred in the Gardens, June 24, 1860.

385. *Querquedula circia* (Linn.). Garganey Teal.
   Hab. Europe.
   a. Male. Presented by Earl Fitzwilliam, Feb. 11, 1858.
   b. Female. Bred in the Gardens, July 19, 1859.

Genus Chaulelasmus.

386. *Chaulelasmus streperus* (Linn.). Common Gadwall.
   Hab. Europe.
   a. Male. b, c, d. Females. Bred in the Gardens, June 20, 1861.
   e, f. Deposited by Mr. Bartlett.

ANATIDÆ.

Genus SPATULA.

387. *Spatula clypeata* (Linn.). Shoveler.
   Hab. Europe.
   a, c. Males. Bred in the Gardens, July 4, 1859.
   b, d. Females. Bred in the Gardens, June 24, 1860.

Subfamily FULIGULINÆ.

Genus FULIGULA.

388. *Fuligula cristata* (Ray). Tufted Duck.
   Hab. Europe.
   a. Male. Purchased, 1862.
   b, c. Females. Purchased, Nov. 22, 1860.
   d, e, f. Hybrids between this species and *Nyroca leucophthalma* (Bechst.), for three or four generations. Bred in the Gardens, June 12, 1861.

389. *Fuligula marila* (Linn.). Scaup Duck.
   Hab. Europe.
   a. Purchased, April 24, 1861.

390. *Fuligula ferina* (Linn.). Red-headed Pochard.
   Hab. Europe.
   a, b. Males. *c, d, e.* Females. Purchased, Nov. 29, 1860.

Genus NYROCA.

391. *Nyroca leucophthalma* (Bechst.). White-eyed or Castaneous Duck.
   Hab. Europe.
   a. Male. Purchased, 1857.

392. *Nyroca brunnea*, Eyton. Brown Duck.
   Hab. South Africa.
   a. Purchased, Nov. 8, 1851.

Genus CLANGULA.

393. *Clangula glaucion* (Linn.). Golden-eye.
   Hab. Europe.
   a, b, c. Purchased, April 23, 1861.

ANATIDÆ.—LARIDÆ.

Genus MERGUS.

394. *Mergus albellus*, Linn. Smew.
 Hab. British Islands.
 a. Purchased, March 14, 1861.

Family LARIDÆ.

Genus LARUS.

395. *Larus marinus*, Linn. Greater Black-backed Gull.
 Hab. British Islands.
 a. Male. Presented by W. H. Leach, Esq., June 3, 1861.
 b. Female. Presented by W. N. Turner, Esq., Oct. 27, 1848.
 c. Deposited by Mr. Bartlett, 1861.

396. *Larus fuscus*, Linn. Lesser Black-backed Gull.
 Hab. British Islands.
 a. Presented by — Berry, Esq., Feb. 24, 1841.
 c. Presented by Mrs. Cotton, 1861.

397. *Larus* ——— ? African Gull.
 Hab. North-eastern Africa.
 a. Purchased, Aug. 3, 1859. From Mogador.

398. *Larus argentatus*, Brünn. Herring Gull.
 Hab. Europe.
 a. Presented by S. Redman, Esq., Feb. 4, 1860.
 b. Presented by — Tomkinson, Esq., Oct. 25, 1861.

399. *Larus ridibundus*, Linn. Black-headed Gull.
 Hab. Europe.
 a. Presented by J. Salmon, Esq., July 25, 1856.
 b. Presented by Dr. Bree, May 1861.
 c. Presented by Alfred Newton, Esq., July 21, 1851.

400. *Larus glaucus*, Linn. Glaucous Gull.
 Hab. Europe.
 a. Purchased, Dec. 23, 1859.
 b. Purchased, Oct. 3, 1860.

Family PELECANIDÆ.

Genus PHALACROCORAX.

401. *Phalacrocorax carbo*, Linn.   Common Cormorant.
   Hab. British Islands.
   a. Purchased, Sept. 26, 1853.   From Egypt.

Genus PELECANUS.

402. *Pelecanus onocrotalus*, Linn.   White Pelican.
   Hab. Europe, Asia, Africa.
   a. Purchased, April 22, 1851.   From Egypt.
   b. Purchased, July 9, 1852.   From Egypt.
   c. Purchased, April 12, 1853.   From Egypt.

403. *Pelecanus crispus*, Bruch.   Crested Pelican.
   Hab. Africa.
   a. Purchased, Sept. 26, 1853.   From Egypt.

404. *Pelecanus fuscus*, Linn.   Brown Pelican.
   Hab. South America.
   a, b. Presented by Capt. Abbott, July 18, 1854.

Family ALCIDÆ.

Genus ALCA.

405. *Alca torda*, Linn.   Razorbill.
   Hab. British coasts.
   a, b. Purchased, May 25, 1862.

Genus FRATERCULA.

406. *Fratercula arctica* (Linn.).   Puffin.
   Hab. British coasts.
   a. Purchased, May 25, 1862.

Genus URIA.

407. *Uria troile* (Linn.).   Common Guillemot.
   Hab. British coasts.
   a, b. Purchased, May 25, 1862.

Class REPTILIA.

Order TESTUDINATA.

Family TESTUDINIDÆ.

Genus TESTUDO.

1. *Testudo græca*, Linn.   European Tortoise.
   Hab. South Europe.
   a. Deposited by Dr. Ogle, July 25, 1860.

Family EMYDIDÆ.

Genus MALACOCLEMMYS.

2. *Malacoclemmys concentrica* (Shaw).   Salt-water Terrapin.
   Hab. North America.
   a, b. Presented by George Cavendish Taylor, Esq., F.Z.S., Dec. 3, 1861.

Genus EMYS.

3. *Emys caspica* (Linn.).   Caspian Terrapin.
   Hab. S. Europe.
   a–e. Purchased, 1861.

4. *Emys laticeps*, J. E. Gray.   Gambian Terrapin.
   Hab. West Africa.
   a. Purchased, 1861.

5. *Emys bennettii*, J. E. Gray.   Bennett's Terrapin.
   Hab. Island of Formosa.
   a, b. Presented by Robert Swinhoe, Esq., C.M.Z.S., April 28, 1862.

6. *Emys guttata*, Schweigh.   Speckled Terrapin.
   Hab. North America.
   a. Presented by George Cavendish Taylor, Esq., F.Z.S., Dec. 3, 1861.

Genus CHELYDRA.

7. *Chelydra serpentina* (Linn.).   Alligator Terrapin.
   Hab. North America.
   a–d. Presented by Arthur Russell, Esq., F.Z.S., June 24, 1860.

Family CHELYDIDÆ.

Genus PELOMEDUSA.

8. *Pelomedusa gehafiæ* (Rüpp.). The Gehafia.
   Hab. Eastern Africa.
   a, b. Purchased.

Genus CHELODINA.

9. *Chelodina oblonga*, J. E. Gray. Oblong Chelodine.
   Hab. West Australia.
   a. Presented by W. Ayshford Sandford, Esq., March 18, 1856.

10. *Chelodina longicollis* (Shaw). Long-necked Chelodine.
    Hab. River Yarra, Australia.
    a. Presented by P. Joske, Esq., Jan. 9, 1861.

Order CROCODILIA.

Family CROCODILIDÆ.

Genus ALLIGATOR.

11. *Alligator mississippiensis* (Daud.). Alligator.
    Hab. Mississippi.
    b. Young. Presented by the Hon. Charles Ellis, Dec. 15, 1861.

Genus JACARE.

12. *Jacare* ——? Caiman.
    Hab. South America.
    a. Young. Received in exchange from the Zoological Gardens, Cologne, June 1861.

Genus CROCODILUS.

13. *Crocodilus americanus*, Schneid. Sharp-nosed Crocodile.
    Hab. West Indies.
    a. Purchased, April 17, 1862.
    b, c. Presented by Capt. Rivett, R.M.S.S. Seine, May 20, 1862.

14. *Crocodilus biporcatus*, Cuv. Indian Crocodile.
    Hab. India.
    a. Purchased, April 26, 1862. From the Hoogly near Calcutta.

Order SAURIA.

Family MONITORIDÆ.

Genus MONITOR.

15. *Monitor niloticus,* Hasselq.   Egyptian Monitor.
    Hab. North Africa.
    a. Purchased, Oct. 22, 1861.

16. *Monitor gouldii,* Schleg.   Australian Monitor.
    Hab. Australia.
    a. Purchased, June 20, 1860.

Family LACERTIDÆ.

Genus LACERTA.

17. *Lacerta viridis,* Linn.   Green Lizard.
    Hab. Island of Jersey.
    a, b. Deposited by the Lord Bishop of Oxford, F.Z.S., Dec. 30, 1861.
    c, d. Presented by the Rev. E. C. Taylor, April 4, 1862.

18. *Lacerta agilis,* Linn.   Sand Lizard.
    Hab. Europe.
    a. Deposited by J. H. Gurney, Esq., M.P., F.Z.S., May 1862. From Dorsetshire.

Family SCINCIDÆ.

Genus TRACHYDOSAURUS.

19. *Trachydosaurus rugosus,* J. E. Gray.   Stump-tailed Lizard.
    Hab. New Holland.
    a. Purchased, 1858.
    b. Presented by the Rev. W. H. Hawker, F.Z.S., April 1862.

Genus CYCLODUS.

20. *Cyclodus gigas* (Bodd.).   Great Cyclodus Lizard.
    Hab. Australia.
    a. Purchased, 1856.

Genus TROPIDOLEPISMA.

21. *Tropidolepisma major*, J. E. Gray.   Ocellated Skink.
    Hab. New South Wales.
    a, b. Presented by Geo. Macleay, Esq., F.Z.S., May 19, 1862.

Family GECCOTIDÆ.
Genus GECCO.

22. *Gecco verus*, Merr.   Indian Gecko.
    Hab. India.
    a, b. Presented by Edward Blyth, Esq., C.M.Z.S., 1861.
    c, d, e. Presented by A. Grote, Esq., C.M.Z.S., 1862.

Family CHAMÆLEONIDÆ.
Genus CHAMÆLEO.

23. *Chamæleo vulgaris*, Daud.   Common Chameleon.
    Hab. North Africa.
    a. Presented by M. Shorto, Esq., Sept. 24, 1861.

24. *Chamæleo pumilus*, Latr.   Dwarf Chameleon.
    Hab. South Africa.
    a. Presented by H.E. Sir George Grey, K.C.B., F.Z.S., Governor of New Zealand, May 26, 1861.

Order OPHIDIA.

Family ERYCIDÆ.
Genus CLOTHONIA.

25. *Clothonia johnii*, J. E. Gray.   Amphisbæna.
    Hab. India.
    a. Presented by the Babu Rajendra Mullick, July 14, 1857.

Family BOIDÆ.
Genus PYTHON.

26. *Python reticulatus* (Schn.).   Ceylonese Python.
    Hab. Ceylon.
    a. Purchased, 1850.

27. *Python sebæ* (Gm.). West-African Python.
   Hab. West Africa.
   a. Purchased, 1855.
   b. Male. Purchased, April 18, 1859.
   c. Female *. Purchased, 1849.

28. *Python regius* (Shaw). Royal Python.
   Hab. West Africa.
   a. Presented, 1859.
   b. Purchased, April 8, 1859.

29. *Python molurus* (Linn.). Indian Python.
   Hab. India.
   a. Presented by James Clarke, Esq.
   b. Purchased, April 22, 1862.

Genus Boa.

30. *Boa constrictor*, Linn. Common Boa.
   Hab. South America.
   a. Presented by Lieut. Richardson, R.A., March 9, 1859.
   b. Presented by G. Furness, Esq., May 23, 1859.
   c. Presented by Hippesley Justins, Esq., July 5, 1861.

Genus Epicrates.

31. *Epicrates cenchris* (Linn.). Ringed Boa.
   Hab. Brazil.
   a. Deposited by Dr. Wucherer of Bahia, C.M.Z.S.

Genus Chilobothrus.

32. *Chilobothrus inornatus*, Dum. Yellow Snake.
   Hab. Jamaica.
   a, b, c. Presented by Dr. Bowerbank, Oct. 17, 1855.

Family NATRICIDÆ.

Genus Tropidonotus.

33. *Tropidonotus quincunciatus* (Schleg.). Common River-Snake.
   Hab. India.
   a, b, c, d. Purchased, July 4, 1861.

---

* This is the specimen which deposited eggs, Jan. 13, 1862.

34. *Tropidonotus viperinus* (Merr.). Viperine Snake.
   Hab. North Africa.
   a–e. Hatched in the Gardens, 1860.

35. *Tropidonotus fasciatus* (Linn.). Mocassin Snake.
   Hab. North America.
   a. Purchased, March 16, 1859.

Family COLUBRIDÆ.

Genus COLUBER.

36. *Coluber guttatus,* Linn. Corn Snake.
   Hab. North America.
   a, b. Purchased, March 16, 1859.

Genus CORYPHODON.

37. *Coryphodon blumenbachii* (Merr.). Indian Rat-Snake.
   Hab. India.
   a–d. Presented by the Babu Rajendra Mullick, July 14, 1857.

Family HERPETODRYADIDÆ.

Genus CYCLOPHIS.

38. *Cyclophis vernalis,* Dekay. Grass Snake.
   Hab. Nova Scotia.
   a–d. Presented by A. Downs, Esq., of Halifax, July 9, 1861.

Genus PHILODRYAS.

39. *Philodryas reinhardtii,* Günther. Reinhardt's Tree-Snake.
   Hab. Brazil.
   a. Deposited by Dr. Wucherer of Bahia, C.M.Z.S.

Family ELAPIDÆ.

Genus HOPLOCEPHALUS.

40. *Hoplocephalus superbus,* Günth. Australian Yellow-bellied Snake.
   Hab. South Australia.
   a. Presented by Edward Wilson, Esq., May 11, 1860.

Genus PSEUDECHIS.

41. *Pseudechis porphyriaca* (Shaw). Black Australian Viper.
    Hab. Australia.
    a, b. Presented by Edward Wilson, Esq., May 11, 1860.

Genus NAIA.

42. *Naia haje* (Linn.). African Cobra.
    Hab. South Africa.
    a. Presented by H.E. Sir George Grey, K.C.B., F.Z.S., Governor of New Zealand, May 26, 1861.

43. *Naia tripudians* (Merr.). Indian Cobra.
    Hab. India.
    a. Purchased, May 24, 1862.

Family VIPERIDÆ.

Genus CENCHRIS.

44. *Cenchris piscivorus*, J. E. Gray. Water Viper.
    Hab. North America.
    a–e. Received in exchange from the Jardin des Plantes, Paris, 1858.

Genus CLOTHO.

45. *Clotho nasicornis* (Shaw). Horned Clotho.
    Hab. West Africa.
    a. Purchased, Jan. 6, 1862.

Order BATRACHIA.

Family BUFONIDÆ.

Genus BUFO.

46. *Bufo pantherinus*, Geoffr. St.-Hil. Pantherine Toad.
    Hab. Tunis.
    a. Presented by Dr. P. L. Sclater, Secretary of the Society, March 24, 1859.

47. *Bufo calamita*, Laur. Natterjack Toad.
    Hab. Cornwall.
    a, b, c. Presented by Dr. Lankester, August 1861.

48. *Bufo vulgaris,* Laur. Common Toad.
Hab. England.
a, b, c. Presented, 1860.

Family RANIDÆ.

Genus RANA.

49. *Rana esculenta,* Linn. Edible Frog.
Hab. Europe.
a, b. Purchased, 1862.

50. *Rana halecina,* Catesby. American Green-Frog.
Hab. Nova Scotia.
a, b. Presented by A. Downs, Esq., C.M.Z.S., July 9, 1861.

51. *Rana mugiens,* Merrem. Bull Frog.
Hab. Nova Scotia.
a, b, c. Presented by A. Downs, Esq., C.M.Z.S., July 9, 1861.

52. *Rana clamata,* Daud. Noisy Frog.
Hab. Nova Scotia.
a. Presented by A. Downs, Esq., C.M.Z.S., July 9, 1861.

Genus HYLA.

53. *Hyla arborea* (Linn.). European Tree-Frog.
Hab. Europe.
a, b, c. Purchased, May 12, 1862.

Genus PELODRYAS.

54. *Pelodryas cæruleus* (White). Australian Tree-Frog.
Hab. New South Wales.
a, b. Presented by Geo. Macleay, Esq., F.Z.S., May 19, 1862.

Family SALAMANDRIDÆ.

Genus SALAMANDRA.

55. *Salamandra maculosa* (Linn.). Spotted Salamander.
Hab. Europe.
a. Presented by Mrs. Hopper, June 27, 1861.

Genus TRITON.

56. *Triton cristatus*, Linn.   Crested Newt.
    Hab. British Islands.
    a, b, c. Presented by T. C. Eyton, Esq., F.Z.S., April 14, 1862.

57. *Triton punctatus*, Dugès.   Smooth Newt.
    Hab. British Islands.
    a, b, c. Presented by T. C. Eyton, Esq., F.Z.S., April 14, 1862.

58. *Triton marmoratus*, Latreille.   Marbled Newt.
    Hab. S. Europe.
    a. Purchased, May 1862.   From Portugal.

Genus AMBLYSTOMA.

59. *Amblystoma luridum*, Baird.   Illinois Salamander.
    Hab. Illinois, U. S. A.
    a, b. Presented by the Smithsonian Institution, Dec. 1857.

Family PROTONOPSIDÆ.

Genus SIEBOLDIA.

60. *Sieboldia maxima* (Schleg.).   Gigantic Salamander.
    Hab. Japan.
    a. Purchased, March 12, 1860.
    b. Deposited by Capt. Taylor, July 2, 1861.

Family PROTEIDÆ.

Genus PROTEUS.

61. *Proteus anguinus* (Shaw).   Proteus.
    Hab. Europe.
    a, b. Presented by Dr. H. Falconer, June 27, 1861.

Family LEPIDOSIRENIDÆ.

Genus PROTOPTERUS.

62. *Protopterus annectans*, Owen.   African Lepidosiren.
    Hab. River Gambia.
    a. Presented by James Thompson, Esq., Oct. 27, 1859.

Class PISCES.

Family GASTEROSTEIDÆ.

Genus GASTEROSTEUS.

1. *Gasterosteus spinachia,* Linn.   Fifteen-spined Stickleback.
Hab. British Seas.
   a, b. Purchased, 1861.

Family PERCIDÆ.

Genus PERCA.

2. *Perca fluviatilis,* Linn.   Common Perch.
Hab. British Islands.
   a–f. Purchased, 1854.

Family TRIGLIDÆ.

Genus COTTUS.

3. *Cottus bubalis,* Euphr.   Long-spined Cottus.
Hab. British Seas.
   a. Purchased, 1861.

Family MUGILIDÆ.

Genus MUGIL.

4. *Mugil chelo,* Cuv.   Thick-lipped Grey Mullet.
Hab. British Seas.
   a–j. Purchased, 1860.

Family BLENNIIDÆ.

Genus CENTRONOTUS.

5. *Centronotus gunnellus* (Linn.).   Spotted Gunnel.
Hab. British Seas.
   a. Purchased, 1861.

H

Genus ZOARCES.

6. *Zoarces viviparus,* Cuv.   Viviparous Blenny.
   Hab. British Seas.
   a, b. Purchased, 1861.

Family GOBIIDÆ.

Genus GOBIUS.

7. *Gobius niger,* Linn.   Black Goby.
   Hab. British Seas.
   a. Purchased, 1861.

8. *Gobius minutus,* Pall.   Freckled Goby.
   Hab. British Seas.
   a, b. Purchased, 1862.

Family LABRIDÆ.

Genus CRENILABRUS.

9. *Crenilabrus cornubicus,* Don.   Goldfinny.
   Hab. British Seas.
   a. Purchased, 1861.

10. *Crenilabrus rupestris,* Selby.   Jago's Goldfinny.
    Hab. British Seas.
    a, b. Purchased, 1861.

Genus LABRUS.

11. *Labrus lineatus,* Don.   Green-streaked Wrasse.
    Hab. British Seas.
    a, b, c. Purchased, 1862.

Family PLEURONECTIDÆ.

Genus PLATESSA.

12. *Platessa flesus* (Linn.).   Flounder.
    Hab. British Seas.
    a. Purchased, 1859.

Family CYPRINIDÆ.

Genus CYPRINUS.

13. *Cyprinus carpio*, Linn.   Common Carp.
    Hab. British Islands.
    a. Purchased, 1854.

14. *Cyprinus gibelio*, Bloch.   Prussian Carp.
    Hab. British Islands.
    a. Purchased, 1862.

15. *Cyprinus auratus*, Linn.   Gold Carp.
    Hab. China.
    a–f. Purchased, 1854.
    g–j. Variety. Presented by Capt. Pope, 1861.

Genus LEUCISCUS.

16. *Leuciscus vulgaris*, Flem.   Common Dace.
    Hab. British Islands.
    a, b, c. Purchased, 1861.

17. *Leuciscus rutilus* (Linn.).   Roach.
    Hab. British Islands.
    a, b, c. Purchased, 1862.

Family SALMONIDÆ.

Genus SALMO.

18. *Salmo fario*, Linn.   Common Trout.
    Hab. British Fresh-waters.
    a. Hatched from ova at Paris in February 1860, and presented by A. Smee, Esq.
    b, c, d. Presented by S. Gurney, Esq., M.P., F.Z.S., May 23, 1862.

Family GADIDÆ.

Genus MOTELLA.

19. *Motella mustela* (Linn.).   Five-bearded Rockling.
    Hab. British Seas.
    a, b. Purchased, 1862.

Family ESOCIDÆ.

Genus Esox.

20. *Esox lucius,* Linn. Pike.
 Hab. British Islands.
 a. Purchased, 1853.

Family MURÆNIDÆ.

Genus CONGER.

21. *Conger vulgaris,* Cuv. Conger Eel.
 Hab. British Islands.
 a. Purchased, 1854.

Family SYNGNATHIDÆ.

Genus SYNGNATHUS.

22. *Syngnathus typhle,* Linn. Deep-nosed Pipe-fish.
 Hab. British Islands.
 a, b. Presented by Dr. J. Salter, F.Z.S., 1861.

23. *Syngnathus ophidion,* Linn. Straight-nosed Pipe-fish.
 Hab. British Seas.
 a, b. Presented by Dr. J. Salter, F.Z.S.

THE END.

www.ingramcontent.com/pod-product-compliance
Lightning Source LLC
Chambersburg PA
CBHW020150170426
43199CB00010B/974